BIOMOLECULES

BIOMOLECULES

BY
AUTON GILL
ISLA PATEL

KNOWLEDGE BAKERS

ISBN: 978-93-9001-371-5

Published by : **Knowledge Bakers**
Pune, India
Email: info@knowledgebakers.com

PREFACE

In the intricate tapestry of life, biomolecules serve as the fundamental building blocks, orchestrating the myriad processes that sustain living organisms. From the elegant simplicity of water molecules to the breathtaking complexity of proteins and nucleic acids, biomolecules encapsulate the essence of life itself, embodying the remarkable beauty and diversity of the natural world. As we embark on this journey of exploration, it is imperative to unravel the mysteries of biomolecules and comprehend their pivotal role in shaping life's myriad forms and functions.

This book, titled "Biomolecules" emerges as a comprehensive guide to understanding the intricate world of biomolecular structures, functions, and interactions. Through collaboration with leading experts and researchers in the field, we aim to provide readers with a holistic view of biomolecules and their significance within the realms of biology, chemistry, and beyond.

As editors of this volume, we have curated a collection of chapters that delve into the diverse classes of biomolecules, including carbohydrates, lipids, proteins, nucleic acids, and more. Each chapter offers a nuanced exploration of the structural properties, functional roles, and biochemical mechanisms underlying these essential molecules, illuminating the intricate interplay between structure and function in living systems.

Furthermore, this book aims to bridge the gap between theory and application, offering readers practical insights into how biomolecules contribute to various biological processes, from metabolism and signaling to genetic expression and disease pathways. Through case studies, experimental techniques, and interdisciplinary perspectives, we strive to empower students, researchers, and practitioners to unlock the secrets of biomolecular complexity and harness their transformative potential in diverse scientific endeavors.

As we embark on this enlightening journey together, we extend our gratitude to the contributors whose expertise and dedication have enriched this volume. We hope that "Biomolecules: Exploring the Building Blocks of Life" serves as a valuable resource and inspires curiosity, discovery, and innovation in the fascinating field of biomolecular science.

Auton Gill
Isla Patel

CONTENTS

CHAPTER-1

CELL: BASIC STRUCTURE AND FUNCTIONS

1.1 CELL

Cells are a basic unit of life, sharing many common mechanisms across different classes of organisms (plants, animals, microorganisms). Between these classes there are also key differences in cell architecture. In a cell, the external bilayer (plasma membrane) and the nuclear membrane are built and maintained by lipids. Stiff protein filaments (microtubules and actin filaments) confer rigidity to the cell's cytoskeleton.The rate of production of proteins (gene regulation) is controlled by many factors, including the physical forces acting on a cell and the cell's mechanical environment.

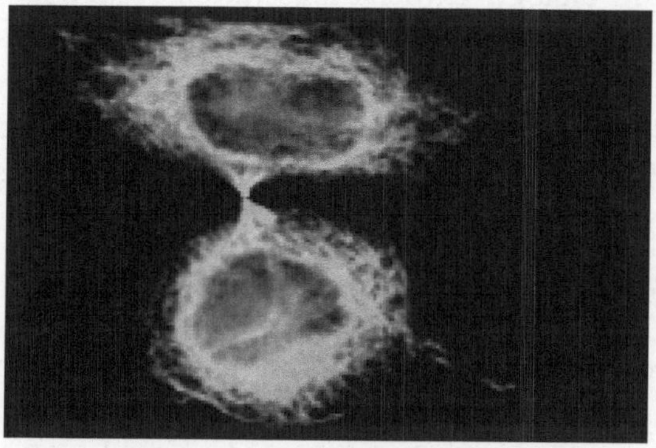

Fig.1: Cell

The cell is a basic unit for life forms. As well as enabling sophisticated control of biochemical processes by providing compartments and regulating chemical fluxes between them, cells also have structural integrity and can exert forces. In the case of multicellular organisms (animals and plants), each cell contributes some mechanical property to the tissue it forms together with other cells. Furthermore, many cells are eliminated during the life of a complex organism (e.g. skin layers in animals), which entails cell division and restructuring of the organisation with neighbours. Some types of cell are actually very motile, moving through tissues (e.g. various immune system cells and some cancer cells). This dynamic aspect is even more obvious during the development of multicellular organisms, when many stages of cell division and migration take place.

1.2 CELL COMPONENTS

A first division of organisms is between those whose cells have within them a nucleus, the structure containing most of the genetic material in the form of DNA, and those whose cells don't. The nucleated cells are called eukaryotic and are found in animals, plants, fungi, protozoa and algae. In contrast, bacteria (and the less common archaea) do not have a nucleus and their DNA is spread throughout the cell. These cells are called prokaryotic. Eukaryotic organisms can be unicellular or multicellular while all prokaryotes are unicellular.

Aside from the distribution of DNA, there are other very obvious differences between eukaryotic and prokaryotic cells. At a coarse level the size, and at the level of molecular detail the chemical detail of proteins used in the two classes of organisms is quite different. (This is what allows certain antibiotics to interfere at the molecular level with essential processes in bacteria, while not affecting the functioning of the cells of an animal.)

Within the nucleated cells there are structural differences between animal and plant cells. A standard biology textbook illustration of an animal eukaryotic cell is shown in Slide 2. In animals, cells can often secrete protein fibres and complex sugar molecules (polysaccharides). These mesh together at the molecular scale, making an elastic material known as extracellular matrix (ECM). Some cell types reside at a particular place in the organism (e.g. the epithelium) while others are motile through the ECM (e.g. immune system cells).

1.3 CELL PROCESSES

Cells regulate their behaviour and their function by controlling the concentration (in some cases it may be small numbers) of different proteins. Some proteins perform a function themselves, whereas others serve to control protein production rates.

This "network" of interactions is analogous to a computing machine, and, indeed, cells can compute. However, the circuit is not hard set. It is not enough to know the genetic code since the cell behaviour will also depend most importantly on the concentration of all proteins, but also on other factors among which are the conformation of the DNA and the composition of lipids in the membranes. There is recent evidence that forces acting on a cell, and more generally the mechanical environment around a cell, can affect the process of gene regulation, maybe one day offering new opportunities to exploit stem cells for regenerative medicine. Before looking at how this might work, it is useful to review very briefly the essential processes that take place in a cell.

Cells contain DNA, which is a sequence of amino-acid bases. The cells in a human all have the same DNA but are clearly very different. About 200 different cell types are classified in a human (bone, skin, blood, muscle, etc.). The differences between cells (biologists would call this the phenotype) are due to each being in a different steady state of protein expression. Proteins are constantly being made and degraded in a cell. The rate of production is finely controlled, involving a sophisticated interplay between binding constants determined by detailed protein structures, and statistical physics

Despite much research in the medical and biological communities, and joined by increasing numbers of physicists and engineers, it is still not fully understood how this fine level of control of the steady state is achieved. This is a key question, the importance of which is clear if we think of stem cells . These are cells, present in all multicellular organisms (at least during some stage of their development from single cell embryos), that are in a state from which they are able to differentiate into any other cell type. Clearly they have the same genetic material as any other cell, and yet they have the unique property of easily becoming "other" cell types. Understanding better how they maintain their special non-differentiated state, and also the cues (which may be chemical or mechanical) that control their differentiation fate, could allow breakthroughs in the treatment of various diseases. The accepted view of this general way in which a cell works is often called the central dogma of molecular cell biology.

Short sequences of DNA containing the information sufficient to describe a protein sequence (a gene) are copied (by a protein machine called RNA polymerase) into RNA. This process is known as transcription.

There are subtle different functions for the RNA, but the main one is for these RNA sequences to be made into proteins. This process, through which the RNA code is read, and "translated" into a sequence of residues to make a specific protein, is called translation and involves a protein machine called the ribosome.

Proteins are the main "workhorses" of a cell: they do things either in isolation or by assembling with multiple units of each other, or of different proteins. The combinations are almost endless. As a general principle, cells need to have different proteins at different times, and also they need to use the energy and materials at their disposal efficiently. For both reasons the production of each type of protein is regulated.

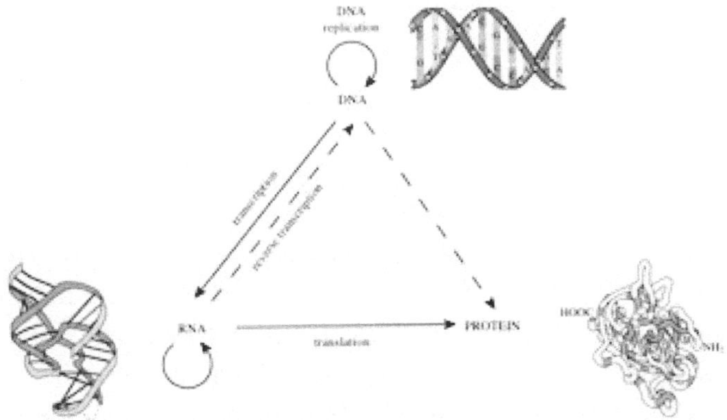

Fig.2: Cell process

This regulation of gene expression happens most importantly at the transcription stage, since this is the first stage and hence it is most efficient to take regulating action there. The main process of regulation involves RNA polymerase binding to DNA in order to start transcription, and this binding affinity can be controlled by certain proteins known as "transcription regulation factors". So by making a few more regulation factor proteins, a cell can tune how many of various other proteins to make.

1.4 CELL TYPES

There are two primary types of cells: eukaryotic cells and prokaryotic cells. Eukaryotic cells are called so because they have a true nucleus. The nucleus, which houses DNA, is contained within a membrane and separated from other cellular structures. Prokaryotic cells however have no true nucleus. DNA in a prokaryotic cell is not separated from the rest of the cell but coiled up in a region called the nucleoid.As organized in the Three Domain System, prokaryotes include Achaeans and bacteria. Eukaryotes include animals, plants, fungi and protists. Typically, eukaryoitc cells are more complex and much larger than prokaryotic cells. On average, prokaryotic cells are about 10 times smaller in diameter than eukaryotic cells. Eukaryotes grow and reproduce through a process called mitosis. In organisms that also reproduce sexually, the reproductive cells are produced by a type of cell division called meiosis.

Most prokaryotes reproduce through a process called binary fission. During binary fission, the single DNA molecule replicates and the original cell is divided into two identical daughter cells. Both eukaryotic and prokaryotic organisms get the energy they need to grow and maintain normal cellular function through cellular respiration. Cellular respiration has three main stages: glycolysis, the citric acid cycle, and electron transport. In eukaryotes, most cellular respiration reactions take place within the mitochondria. In prokaryotes, they occur in the cytoplasm and/or within the cell membrane. Here's a simple visual comparison between a prokaryotic cell and a eukaryotic cell:

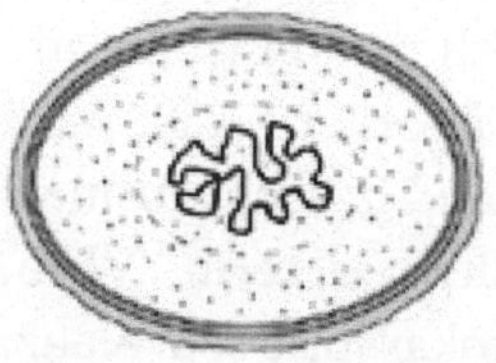

Prokaryotic Cell

(Eukaryotic) Cell

This particular eukaryotic cell happens to be an animal cell, but the cells of plants, fungi and protists are also eukaryotic. Despite their apparent differences, these two cell types have a lot in common. They perform most of the same kinds of functions, and in the same ways. Both are enclosed by plasma membranes, filled with cytoplasm, and loaded with small structures called ribosomes. Both have DNA which carries the archived instructions for operating the cell. And the similarities go far beyond the visible--physiologically they are very similar in many ways. For example, the DNA in the two cell types is precisely the same kind of DNA, and the genetic code for a prokaryotic cell is exactly the same genetic code used in eukaryotic cells.

Some things which seem to be differences aren't. For example, the prokaryotic cell has a cell wall, and this animal cell does not. However, many kinds of eukaryotic cells do have cell walls.

Despite all of these similarities, the differences are also clear. It's pretty obvious from these two little pictures that there are two general categories of difference between these two cell types: size and complexity. Eukaryotic cells are much larger and much more complex than prokaryotic cells. These two observations are not unrelated to each other.

- ☆ If we take a closer look at the comparison of these cells, we see the following differences:
- ☆ Eukaryotic cells have a true nucleus, bound by a double membrane.

Prokaryotic cells have no nucleus. The purpose of the nucleus is to sequester the DNA-related functions of the big eukaryotic cell into a smaller chamber, for the purpose of increased efficiency. This function is unnecessary for the prokaryotic cell, because its much smaller size means that all materials within the cell are relatively close together. Of course, prokaryotic cells do have DNA and DNA functions. Biologists describe the central region of the cell as its "nucleoid" (-oid=similar or imitating), because it's pretty much where the DNA is located. But note that the nucleoid is essentially an imaginary "structure." There is no physical boundary enclosing the nucleoid.

☆ Eukaryotic DNA is linear; prokaryotic DNA is circular (it has no ends).

☆ Eukaryotic DNA is complexed with proteins called "histones," and is organized into chromosomes; prokaryotic DNA is "naked," meaning that it has no histones associated with it, and it is not formed into chromosomes. Though many are sloppy about it, the term "chromosome" does not technically apply to anything in a prokaryotic cell. A eukaryotic cell contains a number of chromosomes; a prokaryotic cell contains only one circular DNA molecule and a varied assortment of much smaller circlets of DNA called "plasmids." The smaller, simpler prokaryotic cell requires far fewer genes to operate than the eukaryotic cell.

☆ Both cell types have many, many ribosomes, but the ribosomes of the eukaryotic cells are larger and more complex than those of the prokaryotic cell. Ribosomes are made out of a special class of RNA molecules (ribosomal RNA, or rRNA) and a specific collection of different proteins. A eukaryotic ribosome is composed of five kinds of rRNA and about eighty kinds of proteins. Prokaryotic ribosomes are composed of only three kinds of rRNA and about fifty kinds of protein.

☆ The cytoplasm of eukaryotic cells is filled with a large, complex collection of organelles, many of them enclosed in their own membranes; the prokaryotic cell contains no membrane-bound organelles which are independent of the plasma membrane. This is a very significant difference, and the source of the vast majority of the greater complexity of the eukaryotic cell. There is much more space within a eukaryotic cell than within a prokaryotic cell, and many of these structures, like the nucleus, increase the efficiency of functions by confining them within smaller spaces within the huge cell, or with communication and movement within the cell.

Examination of these differences is interesting. As mentioned above, they are all associated with larger size and greater complexity. This leads to an important observation. These cells are different from each other. However, they are clearly more alike than different, and they are clearly evolutionarily related to each other. Biologists have no significant doubts about the connection between them. The eukaryotic cell is clearly developed from the prokaryotic cell.

One aspect of that evolutionary connection is particularly interesting. Within eukaryotic cells you find a really fascinating organelle called a mitochondrion. And in plant cells, you'd find an additional family of organelles called plastids, the most famous of which is the renowned chloroplast. Mitochondria (the plural of mitochondrion) and chloroplasts almost certainly have a similar evolutionary origin. Both are pretty clearly the descendants of independent prokaryotic cells, which have taken up permanent residence within other cells through a well-known and very common phenomenon called endosymbiosis.

One structure not shown in our prokaryotic cell is called a mesosome. Not all prokaryotic cells have these. The mesosome is an elaboration of the plasma membrane--a sort of rosette of ruffled membrane intruding into the cell.

This diagram shows a trimmed down prokaryotic cell, including only the plasma membrane and a couple of mesosomes. A mitochondrion is included for comparison:

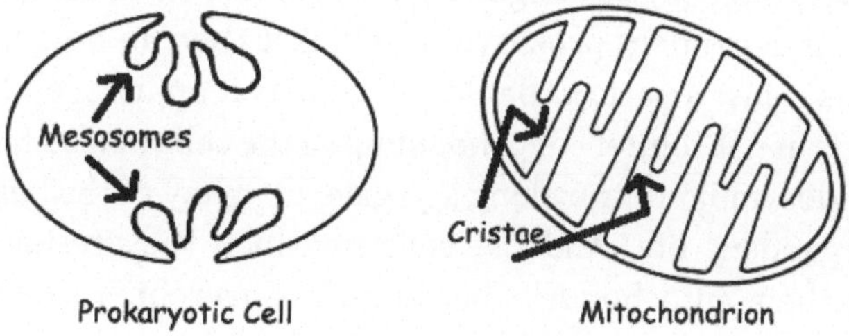

Fig.1: prokaryotic cell

The similarities in appearance between these structures are pretty clear. The mitochondrion is a double-membrane organelle, with a smooth outer membrane and an inner membrane which protrudes into the interior of the mitochondrion in folds called cristae. This membrane is very similar in appearance to the prokaryotic plasma membrane with its mesosomes.

But the similarities are a lot more significant than appearance. Both the mesosomes and the cristae are used for the same function: the aerobic part of aerobic cellular respiration. Cellular respiration is the process by which a cell converts the raw, potential energy of food into biologically useful energy, and there are two general types, anaerobic (not using oxygen) and aerobic (requiring oxygen). In practical terms, the big difference between the two is that aerobic cellular respiration has a much higher energy yield than anaerobic respiration. Aerobic respiration is clearly the evolutionary offspring of anaerobic respiration. In fact, aerobic respiration reallyis anaerobic respiration with additional chemical sequences added on to the end of the process to allow utilization of oxygen (a very common evolutionary pattern--adding new parts to old systems). So it's pretty reasonable of biologists to think that a mitochondrion evolved from a once-independent aerobic prokaryotic cell which entered into an endosymbiotic relationship with a larger, anaerobic cell.

So is there any real evidence that the distant ancestors of mitochondria were independent cells? Quite a lot, actually. And of a very convincing type. Mitochondria (and chloroplasts, for that matter) have their own genetic systems. They have their own DNA, which is not duplicated in the nucleus. That DNA contains a number of the genes which are necessary to make the materials needed for aerobic cellular respiration (or photosynthesis, in the case of the chloroplast). Mitochondrial and chloroplast DNA molecules are naked and circular, like prokaryotic DNA. These organelles also have their own population of ribosomes, which are smaller and simpler than the ribosomes out in the general cytoplasm. Mitochondria and chloroplasts also divide on their own, in a manner similar to the binary fission of prokaryotic cells.

Endosymbiosis is a very widespread phenomenon. The more we look, the more examples we find throughout the kingdoms of life. So, if a mitochondrion is the distant descendent of an independent prokaryotic cell, is it then an organism living inside a larger cell? Or is it just a part of that larger cell? Is it an independent organism or not? Mitochondria are absolutely dependent upon the cells in which they reside. Like any long-time endosymbiont, they long ago gave up many of the basic life processes needed for independent life. And the cells in which they reside are completely dependent upon their mitochondria, because the anaerobic respiration they could do without the mitochondria wouldn't provide nearly enough energy for the cell's needs. In fact, it's very probable that the evolution of big, complex eukaryotic cells wasn't possible until the "invention" of aerobic respiration.

But there are many endosymbiotic relationships in nature which are just as interdependent. For example, no termite could survive without the population of endosymbionts that lives inside its guts, digesting its woody diet for it. And the protists and bacteria that make up that population can't survive outside the termite. Complete interdependency.

Now, the termite and its passengers look a lot more like independent creatures to us than a cell and its mitochondria. But they are actually no more independent of each other. So if we decide that the mitochondrion is just a part of the cell, then don't we have to also decide that the endosymbionts inside the termite's guts are just parts of the termite? If not, how do we justify insisting that there's a difference?

1.4.1 Prokaryotic Cells

Cells that lack a membrane-bound nucleus are called prokaryotes (from the Greek meaning before nuclei). These cells have few internal structures that are distinguishable under a microscope. Cells in the monera kingdom such as bacteria and cyanobacteria (also known as blue-green algae) are prokaryotes.

Prokaryotic cells differ significantly from eukaryotic cells. They don't have a membrane-bound nucleus and instead of having chromosomal DNA, their genetic information is in a circular loop called a plasmid. Bacterial cells are very small, roughly the size of an animal mitochondrion (about 1-2μm in diameter and 10 μm long). Prokaryotic cells feature three major shapes: rod shaped, spherical, and spiral. Instead of going through elaborate replication processes like eukaryotes, bacterial cells divide by binary fission.

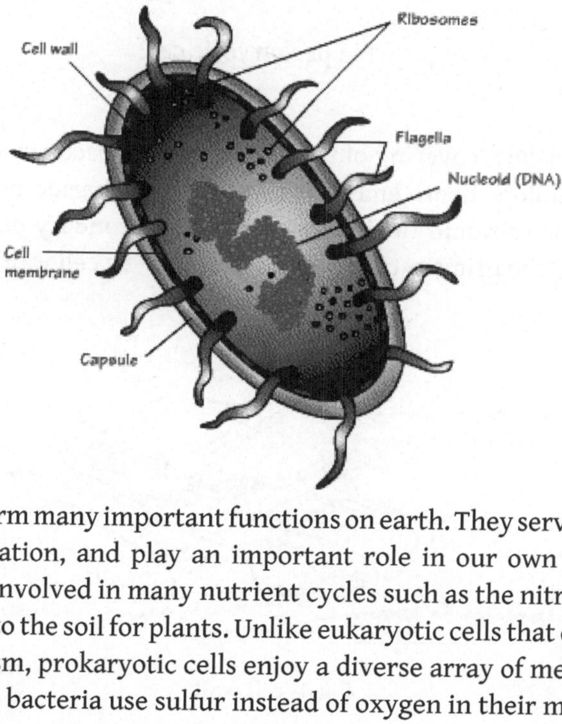

Bacteria perform many important functions on earth. They serve as decomposers, agents of fermentation, and play an important role in our own digestive system. Also, bacteria are involved in many nutrient cycles such as the nitrogen cycle, which restores nitrate into the soil for plants. Unlike eukaryotic cells that depend on oxygen for their metabolism, prokaryotic cells enjoy a diverse array of metabolic functions. For example, some bacteria use sulfur instead of oxygen in their metabolism.

A general prokaryotic cell can be considered to have three architectural regions, though not all the components are always present:

- ☆ A cell envelope consisting of a capsule, a cell wall, and a cell membrane
- ☆ A cytoplasmic region that contains the cell genome (DNA), ribosomes, and various sorts of inclusions

☆ Appendages, sometimes present, called flagella and pili

In A prokaryotic cell, most of the functions of organelles, such as mitochondria, chloroplasts, and the Golgi apparatus, are taken over by the prokaryotic cell membrane.

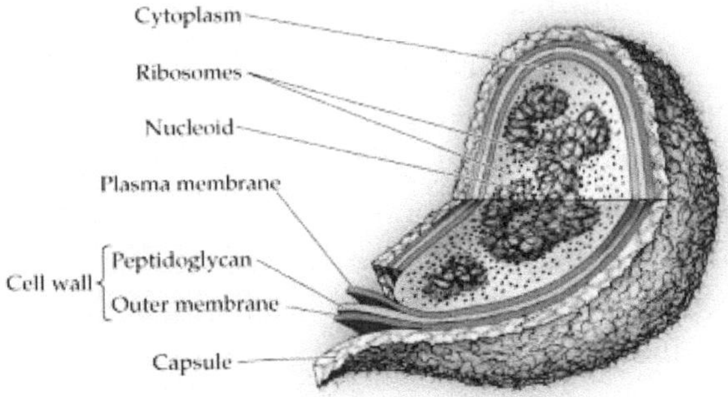

Fig prokaryotic cell

Cell Wall

The rigid, outermost covering of the cells of plants, bacteria, and some protists. It is formed of cellulose fibers embedded in a polysaccharide-protein matrix. The primary cell wall is thin and flexible, whereas the secondary cell wall is stronger and more rigid and the primary constituent of wood. The cell wall surrounds the cell membrane.

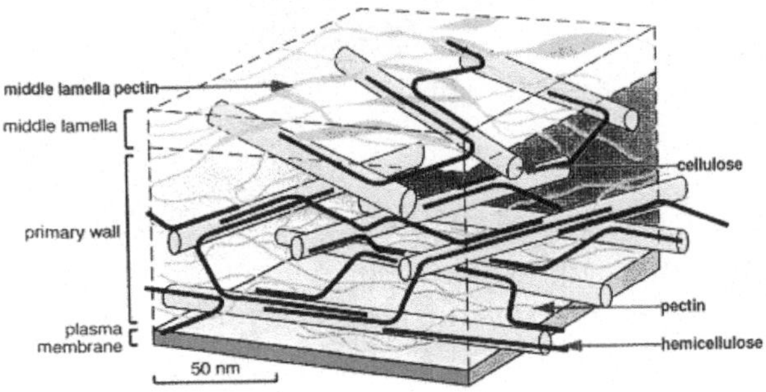

Fig.1: cell wall

Cell membrane

A cell membrane is any membrane found in a living cell. However, unless otherwise specified, it normally refers to the plasma membrane that surrounds the cytoplasm of a cell and forms the cell boundary.

The cell membrane consists of a lipid bilayer with embedded proteins. Depending on the membrane's location and role in the body, lipids can make up anywhere from

20 to 80 percent of the membrane, with the remainder being proteins. Lipids generally give membranes their flexibility. Cholesterol is a type of lipid that, by contrast, helps stiffen the membrane of animal cells but is not found in plant cells.

The cell membrane regulates what enters and leaves the cell, maintains the correct intracellular pH level, and provides a means of separating charges so that the cell can, for example, generate the energy-carrying molecule adenosine triphosphate. Proteins transmit chemical messages into the cell, and they also monitor and maintain the cell's chemical climate. On the outside of cell membranes, attached to some of the proteins and lipids, are chains of sugar molecules that help each cell type do its job.

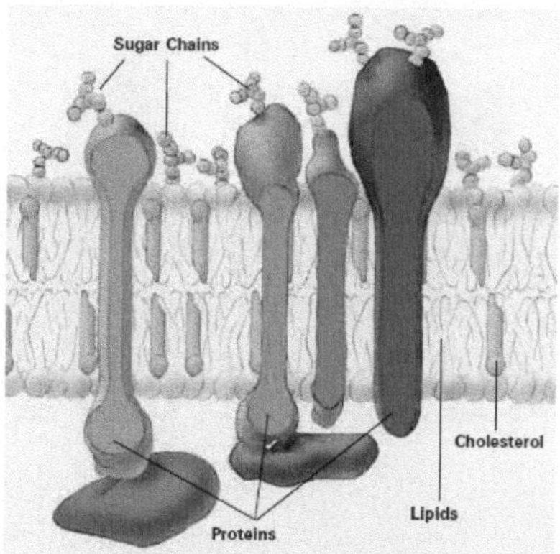

Fig cell membrane

Ribosome

A tiny organelle that is the site of protein synthesis (protein translation) in the living cell. Ribosomes are complex, bead-like structures composed of about 40% protein and 60%ribosomal RNA (rRNA). In eukaryotes, ribosomes are made of four strands of RNA and are often attached to the membranes of the endoplasmic reticulum to form rough ER. In prokaryotes, they are made of three strands of RNA and occur free in the cytoplasm.

Eukaryote ribosomes are produced and assembled in the nucleolus. Three of the four strands are produced there, but one is produced outside the nucleolus and transported inside to complete the ribosome assembly. Ribosomal proteins enter the nucleolus and combine with the four strands to create the two subunits that will make up the completed ribosome. The ribosome units leave the nucleusthrough the nuclear pores and unite once in the cytoplasm. Some ribosomes will remain free-floating in the cytoplasm, creating proteins for the cell's use. Others will attach to the endoplasmic reticulum and produce the proteins that will be "exported" from the cell.

Protein synthesis requires the assistance of two other RNA molecules. Messenger RNA (mRNA) provides instructions from the cellular DNA for building a specific protein. Transfer RNA (tRNA) brings the protein building blocks, amino acids, to the ribosome. Once the protein backbone amino acids are polymerized, the ribosome releases the protein and it is transported to the Golgi apparatus. There, the proteins are completed and released inside or outside the cell.

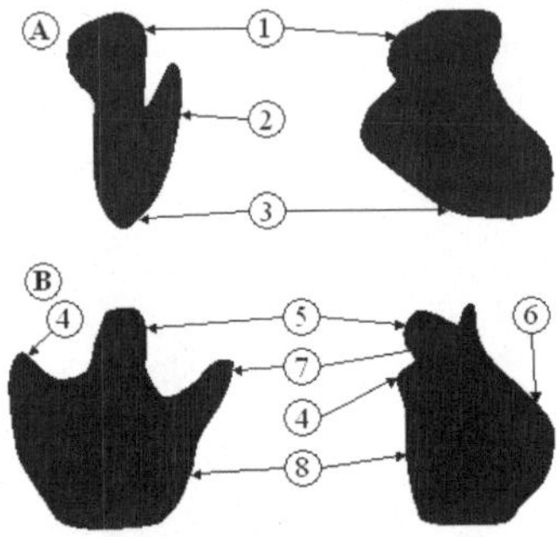

Fig.1:Ribosome structure

Ribosome structure indicating small subunit (A) and large subunit (B) as

- ☆ Head
- ☆ Platform
- ☆ Base
- ☆ Ridge
- ☆ Central protuberance
- ☆ Back
- ☆ Stalk
- ☆ Front

Prokaryotic Cell Division

Prokaryotes are much simpler in their organization than are eukaryotes. There are a great many more organelles in eukaryotes, also more chromosomes. The usual method of prokaryote cell division is termed binary fission. The prokaryotic chromosome is a single DNA molecule that first replicates, then attaches each copy to a different part of the cell membrane. When the cell begins to pull apart, the replicate and original chromosomes are separated. Following cell splitting (cytokinesis), there are then two cells of identical genetic composition (except for the rare chance of a spontaneous mutation).

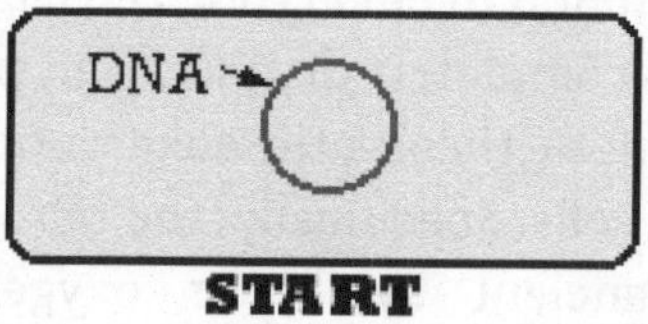

The prokaryote chromosome is much easier to manipulate than the eukaryotic one. We thus know much more about the location of genes and their control in prokaryotes. One consequence of this asexual method of reproduction is that all organisms in a colony are genetic equals. When treating a bacterial disease, a drug that kills one bacteria (of a specific type) will also kill all other members of that clone (colony) it comes in contact with.

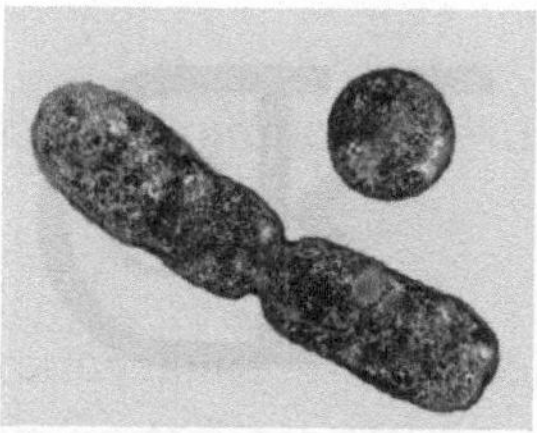

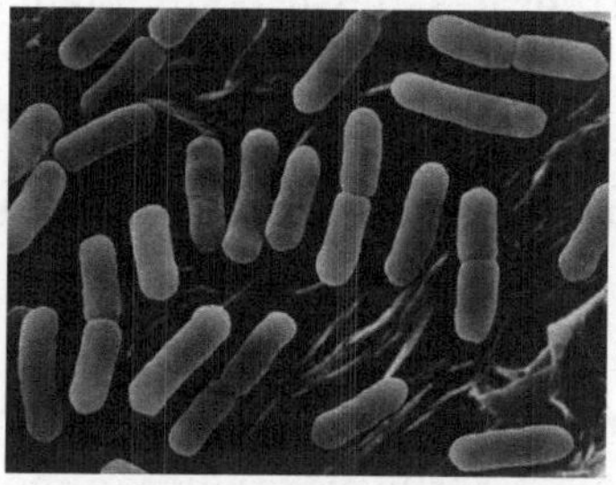

1.4.2 Eukaryotic cell

A cell that contains membrane-bound compartments in which specific metabolic activities take place. Most important among these compartments is the nucleus, which houses the eukaryotic cell's DNA. It is this nucleus that gives the eukaryote – literally, "true nucleus" – its name. Eukaryotic organisms also have other specialized, membrane-bounded structures, called organelles, which are small structures within cells that perform dedicated functions. Eukaryotic cells are typically 10 to 100 micrometers across, or about 10 times the size of prokaryotic cells.

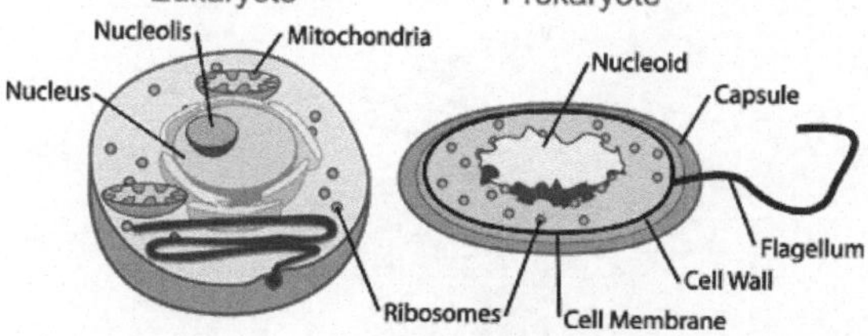

Fig Cells

Origin of Eukaryotic cells

The set of ideas most favored by biologists to explain how eukaryotic cells first came about is called the end symbiotic theory. This theory is able to account well for the fact that two of the organelles found in eukaryotic cells, mitochondria and chloroplasts, have their own DNA that is completely distinct from the DNA housed in the nucleus. According to the endosymbiotic theory, the eukaryotic cell evolved from a symbiotic community of prokaryotic cells. Specifically, the mitochondria and the chloroplasts are what remains of ancient symbiotic oxygen-breathing bacteria and cyan bacteria, respectively, whereas the rest of the cell seems to be derived from an ancestral archaean prokaryote cell.

The origin of the eukaryotic cell was a milestone in the evolution of life. Although eukaryotes use the same genetic code and metabolic processes as prokaryotes, their higher level of organizational complexity has permitted the development of truly multicellular organisms. Without eukaryotes, the world would lack mammals, birds, fish, invertebrates, mushrooms, plants, and complex single-celled organisms.

Structure

All eukaryotic cells have a number of features in common. These include:

Nucleus

The membrane-bounded organelle which, in eukaryotic, houses the chromosomal DNA. It is the place where almost all DNA replication and RNA synthesis occur. The nucleus is spheroidal in shape and separated from the cytoplasm by a double-membrane structure called the nuclear envelope. The nuclear envelope isolates and protects a cell's DNA from various molecules that could accidentally damage its structure or interfere with its processing. During processing, DNA is transcribed, or synthesized, into a special RNA, called mRNA. This mRNA is then transported out of the nucleus, where it is translated into a specific protein molecule. In prokaryotes, DNA processing takes place in the cytoplasm. Inside the nucleus of eukaryotes is the nucleolus.

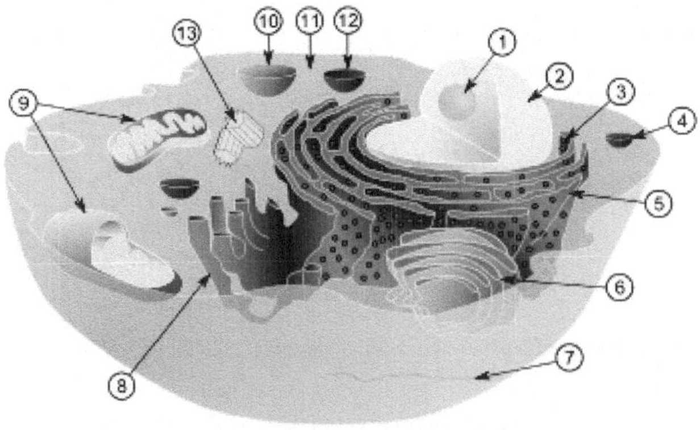

Fig.1: Nucleus

Cytoplasm

The jelly-like matrix that surrounds the nucleus of a cell and is bounded by the cell membrane. It includes the organelles of the cell as well as the sugars, amino acids, and proteins that the cell uses for growth and reproduction. Cytoplasm was formerly referred to as protoplasm. The portion of the cytoplasm that lies outside organelles and includes other insoluble cellular components is called cytosol.

Seen through an electron microscope, the cytoplasm appears as a three-dimensional lattice of thin protein-rich strands known as the microtrabecular lattice (MTL), which serves to interconnect and support the other solid structures in the cytoplasm. Cytoplasm is also the home of the cytoskeleton, a network of cytoplasmic filaments responsible for the movement of the cell and which gives the cell its shape.

The cytoplasm contains dissolved nutrients and salts, helps dissolve waste products, and is an excellent conductor of electricity. It facilitates the movement of materials around the cell by moving and churning through a process called cytoplasmic streaming. The nucleus often flows with the cytoplasm, changing shape as it moves.

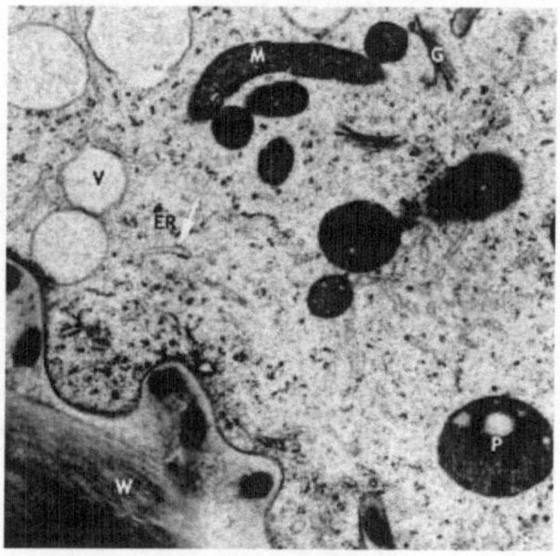

Fig.1:Cytoplasm

Cell (plasma) membrane

A cell membrane is any membrane found in a living cell. However, unless otherwise specified, it normally refers to the plasma membrane that surrounds the cytoplasm of a cell and forms the cell boundary.

The cell membrane consists of a lipid bilayer with embedded proteins. Depending on the membrane's location and role in the body, lipids can make up anywhere from 20 to 80 percent of the membrane, with the remainder being proteins. Lipids generally give membranes their flexibility. Cholesterol is a type of lipid that, by contrast, helps stiffen the membrane of animal cells but is not found in plant cells.

The cell membrane regulates what enters and leaves the cell, maintains the correct intracellular pH level, and provides a means of separating charges so that the cell can, for example, generate the energy-carrying moleculeadenosine triphosphate. Proteins transmit chemical messages into the cell, and they also monitor and maintain the cell's chemical climate. On the outside of cell membranes, attached to some of the proteins and lipids, are chains of sugar molecules that help each cell type do its job.

The cell membrane is both a physical and chemical barrier which defines the boundary between the individual and its environment. Its origin is intimately connected with the origin of life as we know it.

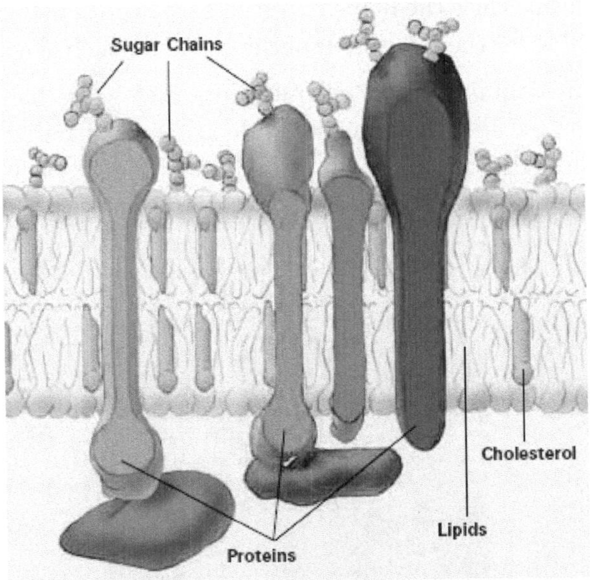

Fig.1: cell membrane

Mitochondria

Mobile, self-replicating organelles that occur in various numbers, shapes, and sizes in the cytoplasm of all eukaryotic cells. Mitochondria contain their own genome that is separate and distinct from the nuclear genome of a cell.

Mitochondria have two functionally distinct membrane systems separated by a space: the outer membrane, which surrounds the whole organelle; and the inner membrane, which is thrown into folds or shelves that project inward. These inward folds are called cristae. The number and shape of cristae in mitochondria differ, depending on the tissue and organism in which they are found, and serve to increase the surface area of the membrane.

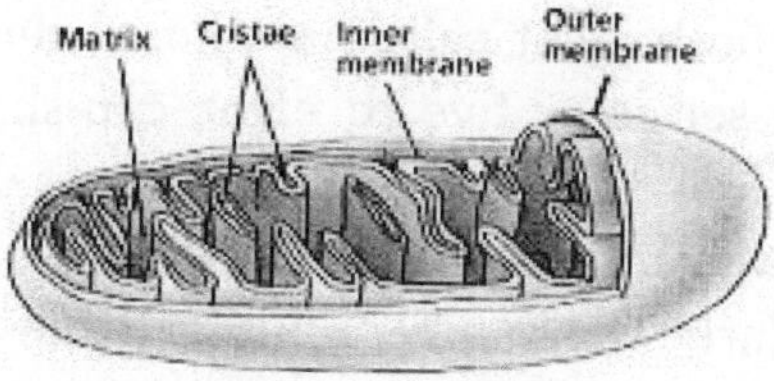

Fig Mitochondria

Endoplasmic reticulum

An extensive network of membranes in the cell that extends from the cell membrane through the cytoplasm to the nuclear envelope. The membranes of the endoplasmic reticulum (ER) surround an inner cavity called the lumen and enclose a series of tubes and flattened membranous areas. The ER membranes actually attach to the cell membrane and the outer membrane of the nuclear envelope as well as the Golgi apparatus in the cytoplasm. The endoplasmic reticulum often makes up more than 10 percent of a cell's total volume.

There are two types of ER. Rough ER has large numbers of ribosomes attached to it and is where new proteins are assembled in the cell. Proteins made on the rough ER's ribosomes end up in other organelles or are sent out of the cell to function elsewhere in the body. A few examples of proteins that leave the cell (called secreted proteins) are antibodies, insulin, digestive enzymes, and manyhormones. (Proteins made on free-floating ribosomes, by contrast, stay in the cytosol.)

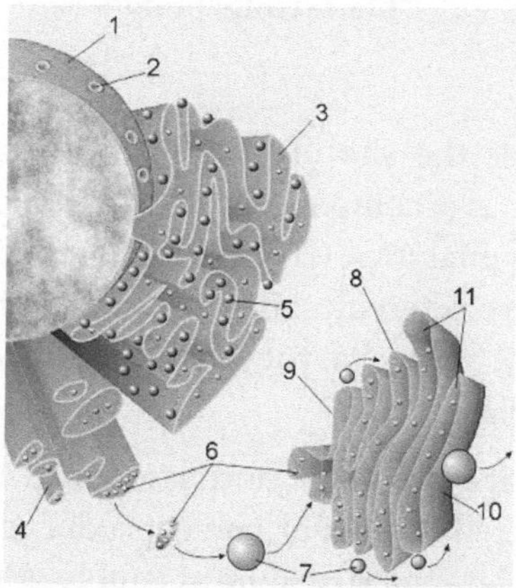

Fig Endoplasmic reticulum

Golgi apparatus

Also known as the Golgi body or Golgi complex, a collection of vesicles and folded membranes in a cell, usually connected to the endoplasmic reticulum (ER). It stores and later transports the proteins manufactured in the endoplasmic reticulum.

Found universally in both plant cells and animal cells, the Golgi apparatus is typically comprised of a series of five to eight cup-shaped, membrane-covered sacs called cisternae that look something like a stack of deflated balloons. In some unicellular flagellates, however, as many as 60 cisternae may combine to make up the Golgi apparatus. Similarly, the number of Golgi bodies in a cell varies according to its function. Animal cells generally contain 10 to 20 Golgi stacks per cell, which are linked into a single complex by tubular connections between cisternae. This complex is usually located close to the cell nucleus.

The Golgi apparatus is particularly well developed in cells that produce secretions, e.g., pancreatic cells producing digestive enzymes.

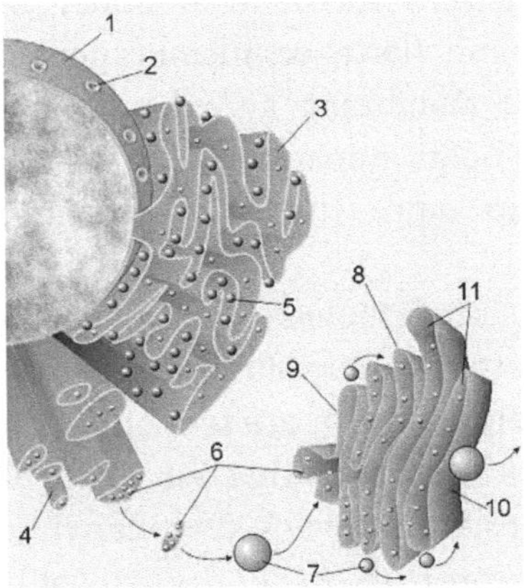

Fig,1: Golgi body

Ribosomes

A tiny organelle that is the site of protein synthesis (protein translation) in the living cell. Ribosomes are complex, bead-like structures composed of about 40% protein and 60% ribosomal RNA (rRNA). In eukaryotes, ribosomes are made of four strands of RNA and are often attached to the membranes of the endoplasmic reticulum to form rough ER. In prokaryotes, they are made of three strands of RNA and occur free in the cytoplasm.

Eukaryote ribosomes are produced and assembled in the nucleolus. Three of the four strands are produced there, but one is produced outside the nucleolus and transported inside to complete the ribosome assembly. Ribosomal proteins enter the nucleolus and combine with the four strands to create the two subunits that will make up the completed ribosome. The ribosome units leave the nucleus through the nuclear pores and unite once in the cytoplasm. Some ribosomes will remain free-floating in the cytoplasm, creating proteins for the cell's use. Others will attach to the endoplasmic reticulum and produce the proteins that will be "exported" from the cell.

Protein synthesis requires the assistance of two other RNA molecules. Messenger RNA (mRNA) provides instructions from the cellular DNA for building a specific protein. Transfer RNA (tRNA) brings the protein building blocks, amino acids, to the ribosome. Once the protein backbone amino acids are polymerized, the ribosome releases the protein and it is transported to the Golgi apparatus. There, the proteins are completed and released inside or outside the cell.

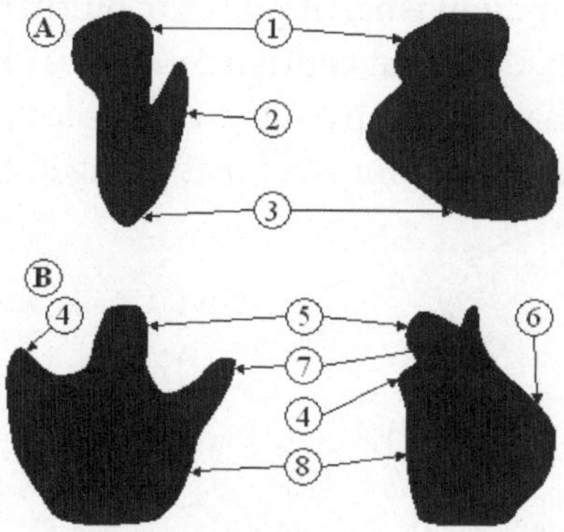

Fig,1: Ribosomes

Eukaryotic Cell Division

Due to their increased numbers of chromosomes, organelles and complexity, eukaryote cell division is more complicated, although the same processes of replication, segregation, and cytokinesis still occur.

Eukaryotic chromosomes occur in the cell in greater numbers than prokaryotic chromosomes. The condensed replicated chromosomes have several points of interest. The kinetochore is the point where microtubules of the spindle apparatus attach. Replicated chromosomes consist of two molecules of DNA (along with their associated histone proteins) known as chromatids. The area where both chromatids are in contact with each other is known as the centromere the kinetochores are on the outer sides of the centromere. Remember that chromosomes are condensed chromatin (DNA plus histone proteins).

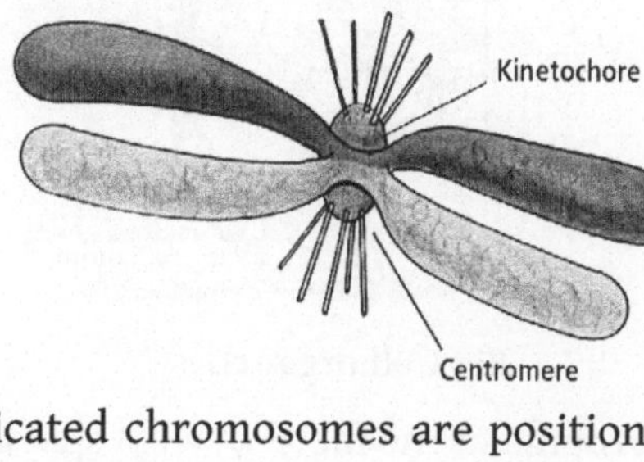

During mitosis replicated chromosomes are positioned near the middle of the cytoplasm and then segregated so that each daughter cell receives a copy of the original DNA (if you start with 46 in the parent cell, you should end up with 46 chromosomes in each daughter cell). To do this cells utilize microtubules (referred to as the spindle apparatus) to "pull" chromosomes into each "cell". The microtubules have the 9+2 arrangement discussed earlier. Animal cells (except for a group of worms known as nematodes) have a centriole. Plants and most other eukaryotic organisms lack centrioles. Prokaryotes, of course, lack spindles and centrioles; the cell membrane assumes this function when it pulls the by-then replicated chromosomes apart during binary fission. Cells that contain centrioles also have a series of smaller microtubules, the aster, that extend from the centrioles to the cell membrane. The aster is thought to serve as a brace for the functioning of the spindle fibers.

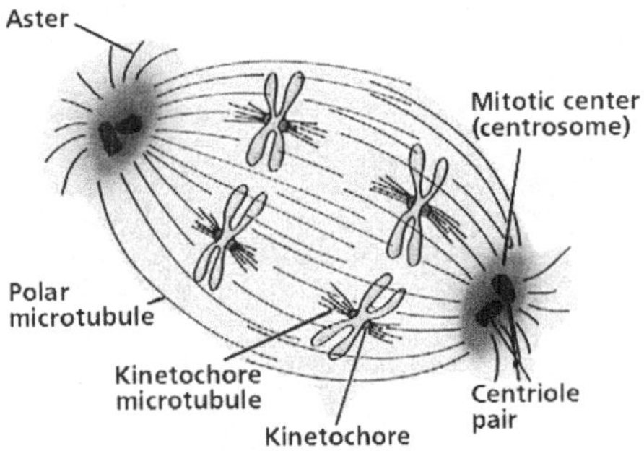

The phases of mitosis are sometimes difficult to separate. Remember that the process is a dynamic one, not the static process displayed of necessity in a textbook.

1.5 CELL ORGANELLES

An organelle is a membrane bound structure found within a cell. Just like cells have membranes to hold everything in, these mini-organs are also bound in a double layer of phospholipids to insulate their little compartments within the larger cells.

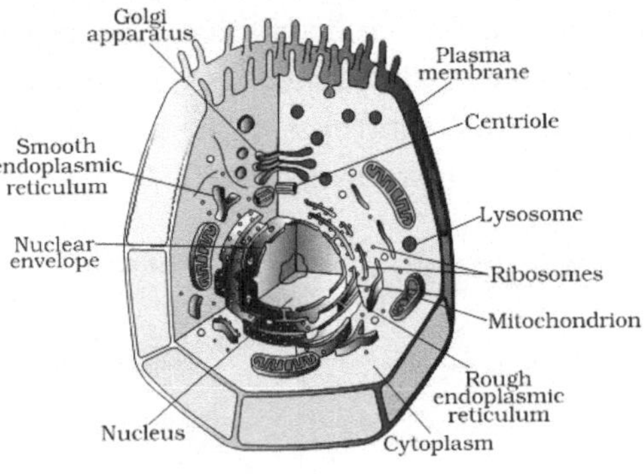

Fig Cell organelles

Organelles are smaller rooms in the factory, with specialized conditions to help these rooms carry out their specific task (like a break room stocked with goodies or a research room with cool gadgets and a special air filter). These organelles are found in the cytoplasm, a viscous liquid found within the cell membrane that houses the organelles and is the location of most of the action happening in a cell. Below is a table of the organelles found in the basic human cell, which we'll be using as our template for this discussion.

Organelle	Function	Factory part
Nucleus	DNA Storage	Room where the blueprints are kept
Mitochondrion	Energy production	Powerplant
Smooth Endoplasmic Reticulum (SER)	Lipid production; Detoxification	Accessory production - makes decorations for the toy, etc.
Rough Endoplasmic Reticulum (RER)	Protein production; in particular for export out of the cell	Primary production line - makes the toys
Golgi apparatus	Protein modification and export	Shipping department
Peroxisome	Lipid Destruction; contains oxidative enzymes	Security and waste removal
Lysosome	Protein destruction	Recycling and security

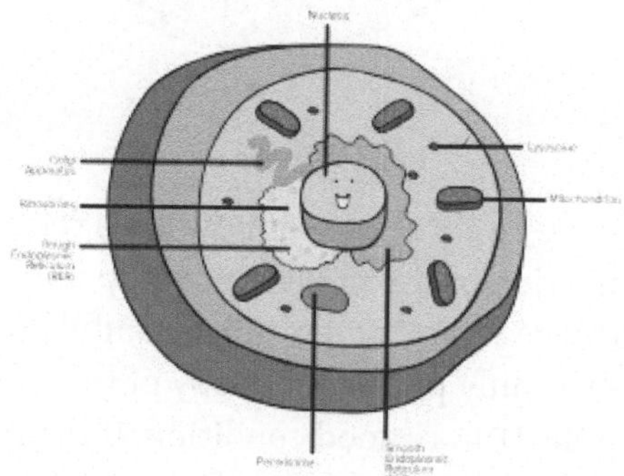

Fig.1: Diagram of cell membrane bound organelles

1.5.1 Nucleus

Our DNA has the blueprints for every protein in our body, all packaged into a neat double helix. The processes to transform DNA into proteins are known as transcription and translation, and happen in different compartments within the cell. The first step, transcription, happens in the nucleus, which holds our DNA. A membrane called the nuclear envelope surrounds the nucleus, and its job is to create a room within the cell to both protect the genetic information and to house all the molecules that are involved in processing and protecting that info. This membrane is actually a set of two lipid bilayers, so there are four sheets of lipids separating the inside of the nucleus from the cytoplasm. The space between the two bilayers is known as the perinuclear space.

Though part of the function of the nucleus is to separate the DNA from the rest of the cell, molecules must still be able to move in and out (e.g., RNA). Proteins channels known as nuclear pores form holes in the nuclear envelope. The nucleus itself is filled with liquid (called nucleoplasm) and is similar in structure and function to cytoplasm. It is here within the nucleoplasm where chromosomes (tightly packed strands of DNA containing all our blueprints) are found.

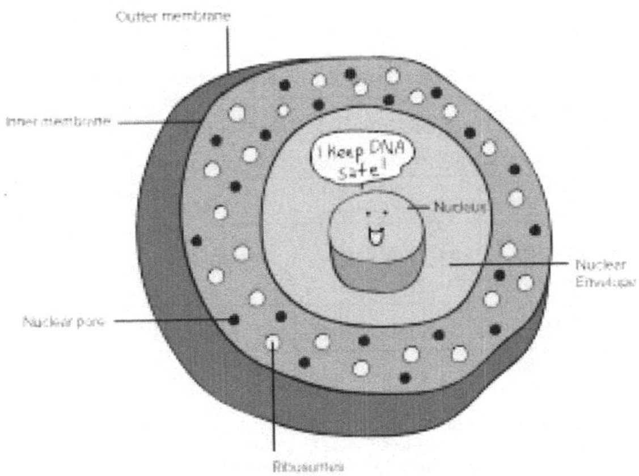

Fig.1: Nucleus

A nucleus has interesting implications for how a cell responds to its environment. Thanks to the added protection of the nuclear envelope, the DNA is a little bit more secure from enzymes, pathogens, and potentially harmful products of fat and protein metabolism. Since this is the only permanent copy of the instructions the cell has, it is very important to keep the DNA in good condition. If the DNA was not sequestered away, it would be vulnerable to damage by the aforementioned dangers, which would then lead to defective protein production. Imagine a giant hole or coffee stain in the blueprint for your toy - all of a sudden you don't have either enough or the right information to make a critical piece of the toy.

The nuclear envelope also keeps molecules responsible for DNA transcription and repair close to the DNA itself - otherwise those molecules would diffuse across the entire cell and it would take a lot more work and luck to get anything done! While transcription (making a complementary strand of RNA from DNA) is completed within the nucleus, translation (making protein from RNA instructions) takes place in the cytoplasm. If there was no barrier between the transcription and translation machineries, it's possible that poorly-made or unfinished RNA would get turned into poorly made and potentially dangerous proteins. Before an RNA can exit the nucleus to be translated, it must get special modifications, in the form of a cap and tail at either end of the molecule, that act as a stamp of approval to let the cell know this piece of RNA is complete and properly made.

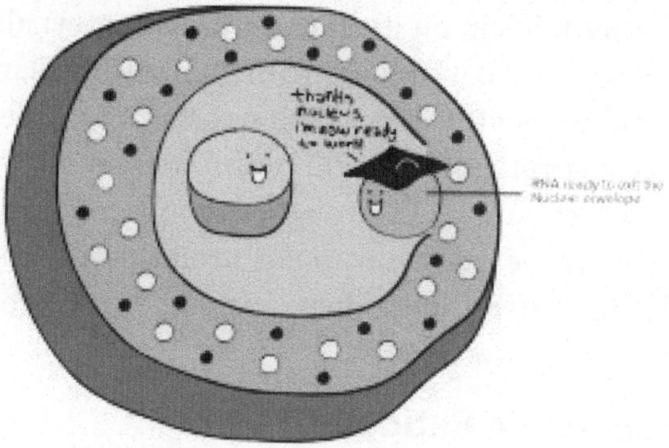

Fig.1: mRNA leaving nucleus and entering cytoplasm.

1.5.2 Nucleolus

Within the nucleus is a small subspace known as the nucleolus. It is not bound by a membrane, so it is not an organelle. This space forms near the part of DNA with instructions for making ribosomes, the molecules responsible for making proteins. Ribosomes are assembled in the nucleolus, and exit the nucleus with nuclear pores. In our analogy, the robots making our product are made in a special corner of the blueprint room, before being released to the factory.

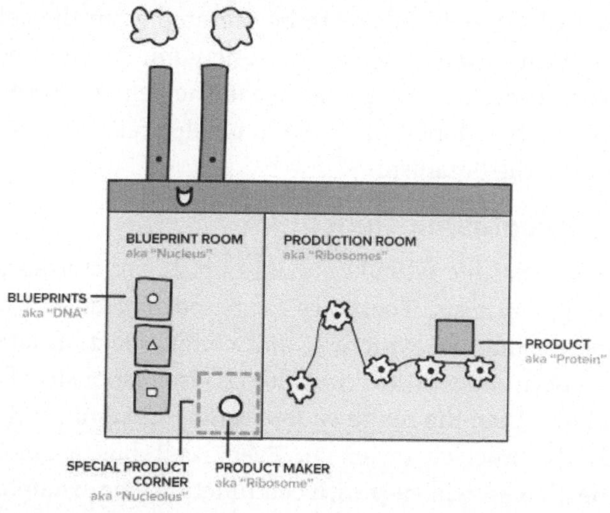

Fig.1: Cell as a factory where cell membrane is shown as factory walls.

1.5.3 Endoplasmic Reticulum

Endoplasmic means inside (endo) the cytoplasm (plasm). Reticulum comes from the Latin word for net. Basically, an endoplasmic reticulum is a plasma membrane found inside the cell that folds in on itself to create an internal space known as the lumen. This lumen is actually continuous with the perinuclear space, so we know the endoplasmic reticulum is attached to the nuclear envelope. There are actually two different endoplasmic reticuli in a cell: the smooth endoplasmic reticulum and the rough endoplasmic reticulum. The rough endoplasmic reticulum is the site of protein production (where we make our major product - the toy) while the smooth endoplasmic reticulum is where lipids (fats) are made (accessories for the toy, but not the central product of the factory).

1.5.4 Rough Endoplasmic Reticulum

The rough endoplasmic reticulum is so-called because its surface is studded with ribosomes, the molecules in charge of protein production. When a ribosome finds a specific RNA segment, that segment may tell the ribosome to travel to the rough endoplasmic reticulum and embed itself. The protein created from this segment will find itself inside the lumen of the rough endoplasmic reticulum, where it folds and is tagged with a (usually carbohydrate) molecule in a process known as glycosylation that marks the protein for transport to the Golgi apparatus. The rough endoplasmic reticulum is continuous with the nuclear envelope, and looks like a series of canals near the nucleus. Proteins made in the rough endoplasmic reticulum as destined to either be a part of a membrane, or to be secreted from the cell membrane out of the cell. Without an rough endoplasmic reticulum, it would be a lot harder to distinguish between proteins that should leave the cell, and proteins that should remain. Thus, the rough endoplasmic reticulum helps cells specialize and allows for greater complexity in the organism.

1.5.5 Smooth Endoplasmic Reticulum

The smooth endoplasmic reticulum makes lipids and steroids, instead of being involved in protein synthesis. These are fat-based molecules that are important in energy storage, membrane structure, and communication (steroids can act as hormones). The smooth endoplasmic reticulum is also responsible for detoxifying the cell. It is more tubular than the rough endoplasmic reticulum, and is not necessarily continuous with the nuclear envelope. Every cell has a smooth endoplasmic reticulum, but the amount will vary with cell function. For example, the liver, which is responsible for most of the body's detoxification, has a larger amount of smooth endoplasmic reticulum.

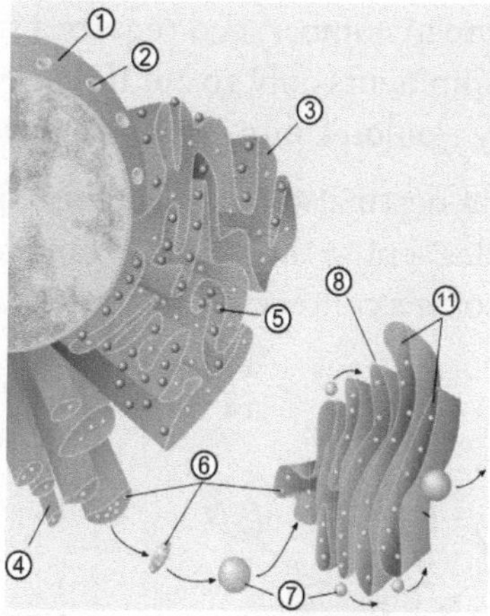

Fig.1: Structure of endoplasmic reticulum, golgi apparatus, smooth endoplasmic reticulum.

Golgi apparatus

We mentioned the Golgi apparatus earlier when we discussed the production of proteins in the rough endoplasmic reticulum. If the smooth and rough endoplasmic reticula are how we make our product, the Golgi is the mailroom that sends our product to customers . It is responsible for packing proteins from the rough endoplasmic reticulum into membrane-bound vesicles (tiny compartments of lipid bilayer that store molecules) which then translocate to the cell membrane. At the cell membrane, the vesicles can fuse with the larger lipid bilayer, causing the vesicle contents to either become part of the cell membrane or be released to the outside.

Different molecules actually have different fates upon entering the Golgi. This determination is done by tagging the proteins with special sugar molecules that act as a shipping label for the protein. The shipping department identifies the molecule and sets it on one of 4 paths:

☆ **Cytosol:** the proteins that enter the Golgi by mistake are sent back into the cytosol (imagine the barcode scanning wrong and the item being returned).

☆ **Cell membrane:** proteins destined for the cell membrane are processed continuously. Once the vesicle is made, it moves to the cell membrane and fuses with it. Molecules in this pathway are often protein channels which allow molecules into or out of the cell, or cell identifiers which project into the extracellular space and act like a name tag for the cell.

☆ **Secretion:** some proteins are meant to be secreted from the cell to act on other parts of the body. Before these vesicles can fuse with the cell membrane, they must accumulate in number, and require a special chemical signal to be released. This way shipments only go out if they're worth the cost of sending them (you generally wouldn't ship just one toy and expect to profit).

☆ **Lysosome:** The final destination for proteins coming through the Golgi is the lysosome. Vesicles sent to this acidic organelle contain enzymes that will hydrolyze the lysosome's content.

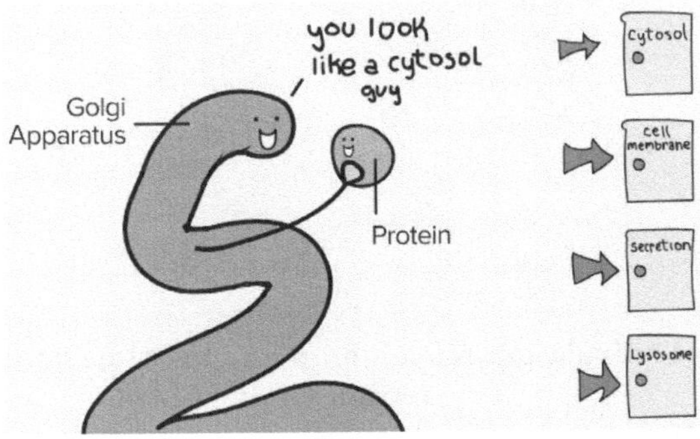

Fig.1: Golgi apparatus sorting proteins into four paths

Lysosome

The lysosome is the cell's recycling center. These organelles are spheres full of enzymes ready to hydrolyze (chop up the chemical bonds of) whatever substance crosses the membrane, so the cell can reuse the raw material. These disposal enzymes only function properly in environments with a pH of 5, two orders of magnitude more acidic than the cell's internal pH of 7. Lysosomal proteins only being active in an acidic environment acts as safety mechanism for the rest of the cell - if the lysosome were to somehow leak or burst, the degradative enzymes would inactivate before they chopped up proteins the cell still needed.

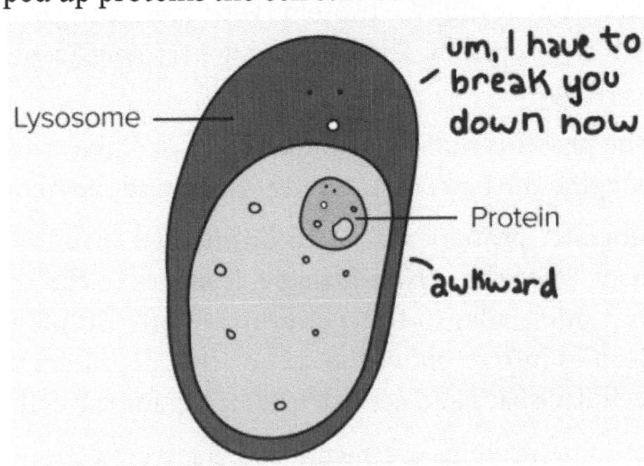

Fig.1: lysosome breaking down protein

Peroxisome

Like the lysosome, the peroxisome is a spherical organelle responsible for destroying its contents. Unlike the lysosome, which mostly degrades proteins, the peroxisome is the site of fatty acid breakdown. It also protects the cell from reactive oxygen species (ROS) molecules which could seriously damage the cell. ROSs are molecules like oxygen ions or peroxides that are created as a byproduct of normal cellular metabolism, but also by radiation, tobacco, and drugs. They cause what is known as oxidative stress in the cell by reacting with and damaging DNA and lipid-based molecules like cell membranes. These ROSs are the reason we need antioxidants in our diet.

Mitochondria

Just like a factory can't run without electricity, a cell can't run without energy. ATP (adenosine triphosphate) is the energy currency of the cell, and is produced in a process known as cellular respiration. Though the process begins in the cytoplasm, the bulk of the energy produced comes from later steps that take place in the mitochondria.

Like we saw with the nuclear envelope, there are actually two lipid bilayers that separate the mitochondrial contents from the cytoplasm. We refer to them as the inner and outer mitochondrial membranes. If we cross both membranes we end up in the matrix, where pyruvate is sent after it is created from the breakdown of glucose (this is step 1 of cellular respiration, known as glycolysis).The space between the two membranes is called the intermembrane space, and it has a low pH (is acidic) because the electron transport chain embedded in the inner membrane pumps protons (H+) into it. Energy to make ATP comes from protons moving back into the matrix down their gradient from the intermembrane space.

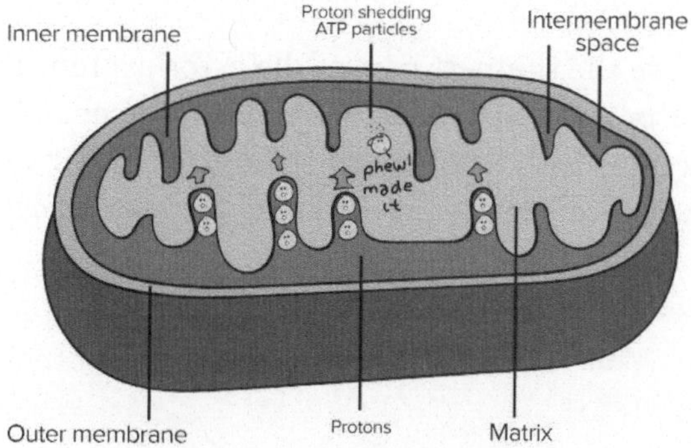

Fig.1: Parts of mitochondria.

Mitochondria are also somewhat unique in that they are self-replicating and have their own DNA, almost as if they were a completely separate cell. The prevailing theory, known as the endosymbiotic theory, is that eukaryotes were first formed by large prokaryotic cells engulfing smaller cells that looked a lot like mitochondria (and chloroplasts, more on them later). Instead of being digested, the engulfed cells remained intact and the arrangement turned out to be advantageous to both cells, which created a symbiotic relationship.

Cytoskeleton

Within the cytoplasm there is network of protein fibers known as the cytoskeleton. This structure is responsible for both cell movement and stability. The major components of the cytoskeleton are microtubules, intermediate filaments, and microfilaments.

Microtubules

Microtubules are small tubes made from the protein tubulin. These tubules are found in cilia and flagella, structures involved in cell movement. They also help provide pathways for secretory vesicles to move through the cell, and are even involved in cell division as they are a part of the mitotic spindle, which pulls homologous chromosomes apart.

Intermediate Filaments

Smaller than the microtubules, but larger than the microfilaments, the intermediate filaments are made of a variety of proteins such as keratin and/ or neurofilament. They are very stable, and help provide structure to the nuclear envelope and anchor organelles.

Microfilaments

Microfilaments are the thinnest part of the cytoskeleton, and are made of actin [a highly-conserved protein that is actually the most abundant protein in most eukaryotic cells]. Actin is both flexible and strong, making it a useful protein in cell movement. In the heart, contraction is mediated through an actin-myosin system.

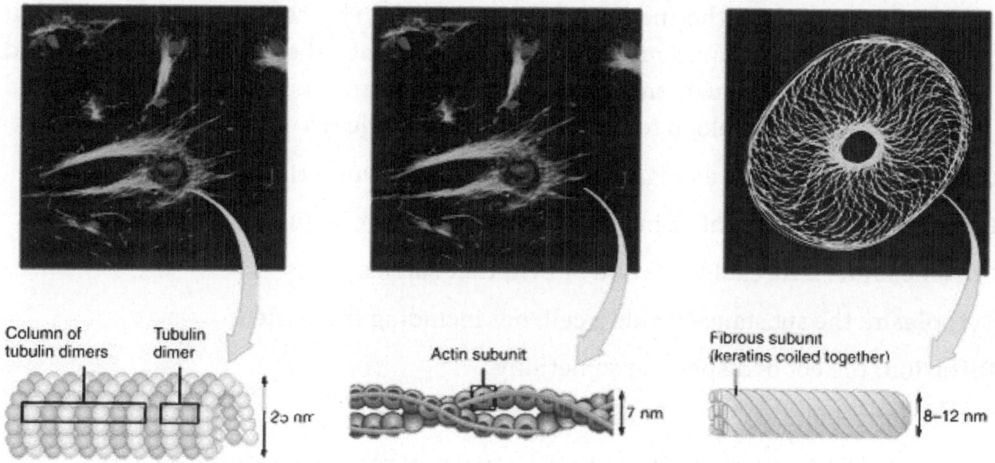

Fig.1: microtubules, microfilaments, intermediate fibers

Plants and Platelets

So far we've covered basic organelles found in a eukaryotic cell. However, not every cell has each of these organelles, and some cells have organelles we haven't discussed. For example, plant cells have chloroplasts, organelles that resemble mitochondria and are responsible for turning sunlight into useful energy for the cell (this is like factories that are powered by energy they collect via solar panels). On the other hand, platelets, blood cells responsible for clotting, have no nucleus and are in fact just fragments of cytoplasm contained within a cell membrane.

Eukaryotes vs Bacteria vs Archaea

It is also important to keep in mind that organelles are found only in eukaryotes, one of the three major cell divisions. The other two major divisions, Bacteria and Archaea are known as prokaryotes, and have no membrane bound organelles within.

GLOSSARY

ATP: a nucleotide derived from adenosine that occurs in muscle tissue; the major source of energy for cellular reactions

Cell: the basic structural and functional unit of all organisms

Cell Membrane: a thin membrane enclosing the cytoplasm of a cell

Cell Wall: a rigid layer of polysaccharides enclosing a plant membrane

Cellular Respiration: the metabolic processes whereby certain organisms obtain energy from organic molecules; processes that take place in the cells and tissues during which energy is released and carbon dioxide is produced and absorbed by the blood to be transported to the lungs

Chlorophyll: any of green pigments found in photosynthetic organisms

Chloroplast: plastid containing chlorophyll and other pigments

Chromosome: a threadlike strand of DNA that carries genes

Cytoplasm: the substance inside a cell, not including the nucleus

Diffusion: the act of dispersing something

DNA: a long linear polymer found in the nucleus of a cell

Mitochondrion: an organelle containing enzymes responsible for producing energy

Mitosis: the process by which a cell divides into two smaller cells

Nucleus: a part of the cell responsible for growth and reproduction

Organelle: a specialized part of a cell; analogous to an organ

Osmosis: diffusion of molecules through a semipermeable membrane

Permeable: allowing fluids or gases to pass or diffuse through

Photosynthesis: synthesis of compounds in plants aided by radiant energy

Ribosome: a particle in a cell that helps synthesize proteins

RNA: a long linear polymer of nucleotides found in the nucleus but mainly in the cytoplasm of a cell where it is associated with microsomes; it transmits genetic information from DNA to the cytoplasm and controls certain chemical processes in the cell

Semipermeable Membrane: a membrane that allows some molecules to pass through but not others

Vacuole: a tiny cavity filled with fluid in the cytoplasm of a cell

CHAPTER-2

CELL MEMBRANE AND TRANSPORT

2.1 CELL MEMBRANE

The cell membrane is an extremely pliable structure composed primarily of back-to-back phospholipids (a "bilayer"). Cholesterol is also present, which contributes to the fluidity of the membrane, and there are various proteins embedded within the membrane that have a variety of functions.

A single phospholipid molecule has a phosphate group on one end, called the "head," and two side-by-side chains of fatty acids that make up the lipid tails. The phosphate group is negatively charged, making the head polar and hydrophilic—or "water loving." A hydrophilic molecule (or region of a molecule) is one that is attracted to water. The phosphate heads are thus attracted to the water molecules of both the extracellular and intracellular environments.

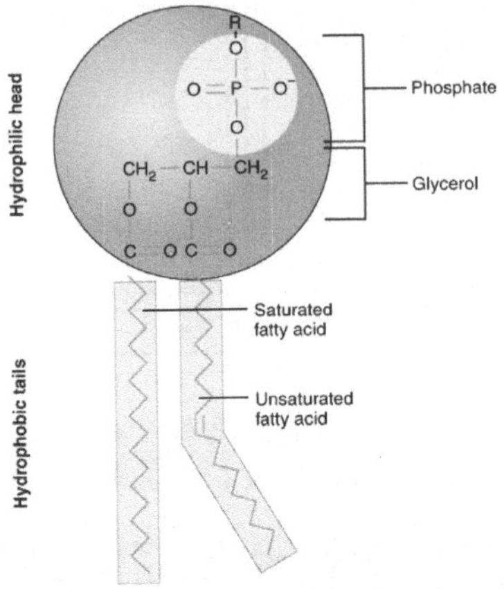

Fig.1: Phospholipid Structure

The lipid tails, on the other hand, are uncharged, or nonpolar, and are hydrophobic—or "water fearing." A hydrophobic molecule (or region of a molecule) repels and is repelled by water. Some lipid tails consist of saturated fatty acids and some contain unsaturated fatty acids. This combination adds to the fluidity of the tails that are constantly in motion. Phospholipids are thus amphipathic molecules. An amphipathic molecule is one that contains both a hydrophilic and a hydrophobic region. In fact, soap works to remove oil and grease stains because it has amphipathic properties. The hydrophilic portion can dissolve in water while the hydrophobic portion can trap grease in micelles that then can be washed away.

The cell membrane consists of two adjacent layers of phospholipids. The lipid tails of one layer face the lipid tails of the other layer, meeting at the interface of the two layers. The phospholipid heads face outward, one layer exposed to the interior of the cell and one layer exposed to the exterior. Because the phosphate groups are polar and hydrophilic, they are attracted to water in the intracellular fluid. Intracellular fluid (ICF) is the fluid interior of the cell. The phosphate groups are also attracted to the extracellular fluid. Extracellular fluid (ECF) is the fluid environment outside the enclosure of the cell membrane.

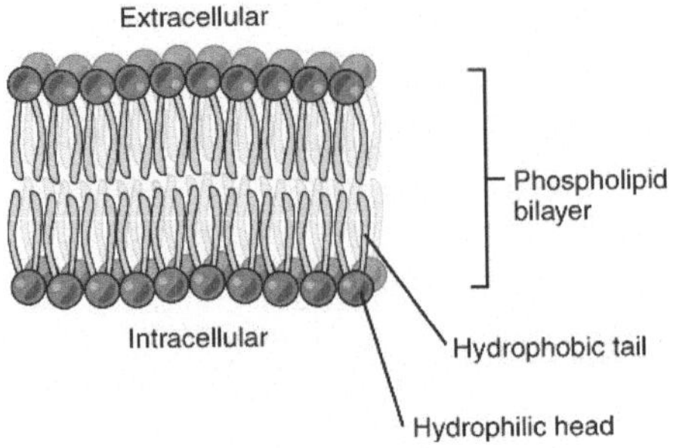

Fig.1: Phospolipid Bilayer

Interstitial fluid (IF) is the term given to extracellular fluid not contained within blood vessels. Because the lipid tails are hydrophobic, they meet in the inner region of the membrane, excluding watery intracellular and extracellular fluid from this space. The cell membrane has many proteins, as well as other lipids (such as cholesterol), that are associated with the phospholipid bilayer. An important feature of the membrane is that it remains fluid; the lipids and proteins in the cell membrane are not rigidly locked in place.

2.1.1 Membrane Proteins

The lipid bilayer forms the basis of the cell membrane, but it is peppered throughout with various proteins. Two different types of proteins that are commonly associated with the cell membrane are the integral proteins and peripheral protein. As its name suggests, an integral protein is a protein that is embedded in the membrane. A channel protein is an example of an integral protein that selectively allows particular materials, such as certain ions, to pass into or out of the cell.

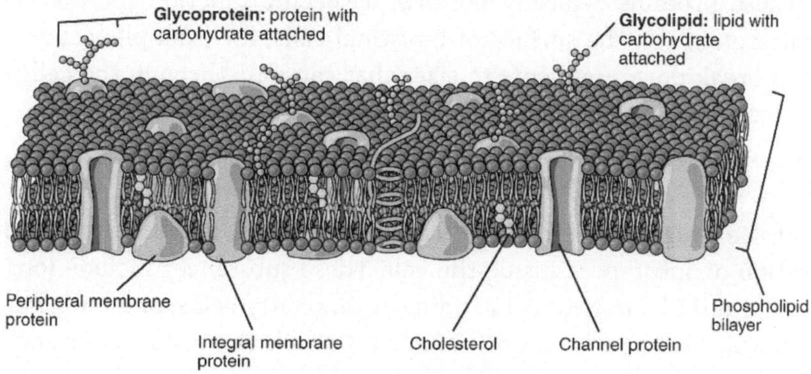

Fig.1: Cell Membrane

Another important group of integral proteins are cell recognition proteins, which serve to mark a cell's identity so that it can be recognized by other cells. A receptor is a type of recognition protein that can selectively bind a specific molecule outside the cell, and this binding induces a chemical reaction within the cell. A ligand is the specific molecule that binds to and activates a receptor. Some integral proteins serve dual roles as both a receptor and an ion channel. One example of a receptor-ligand interaction is the receptors on nerve cells that bind neurotransmitters, such as dopamine. When a dopamine molecule binds to a dopamine receptor protein, a channel within the transmembrane protein opens to allow certain ions to flow into the cell.

Some integral membrane proteins are glycoproteins. A glycoprotein is a protein that has carbohydrate molecules attached, which extend into the extracellular matrix. The attached carbohydrate tags on glycoproteins aid in cell recognition. The carbohydrates that extend from membrane proteins and even from some membrane lipids collectively form the glycocalyx. The glycocalyx is a fuzzy-appearing coating around the cell formed from glycoproteins and other carbohydrates attached to the cell membrane. The glycocalyx can have various roles. For example, it may have molecules that allow the cell to bind to another cell, it may contain receptors for hormones, or it might have enzymes to break down nutrients. The glycocalyces found in a person's body are products of that person's genetic makeup. They give each of the individual's trillions of cells the "identity" of belonging in the person's body. This identity is the primary way that a person's immune defense cells "know" not to attack the person's own body cells, but it also is the reason organs donated by another person might be rejected.

Peripheral proteins are typically found on the inner or outer surface of the lipid bilayer but can also be attached to the internal or external surface of an integral protein. These proteins typically perform a specific function for the cell. Some peripheral proteins on the surface of intestinal cells, for example, act as digestive enzymes to break down nutrients to sizes that can pass through the cells and into the bloodstream.

2.2 TRANSPORT ACROSS CELL MEMBRANE

One of the great wonders of the cell membrane is its ability to regulate the concentration of substances inside the cell. These substances include ions such as Ca^{++}, Na^{+}, K^{+}, and Cl^{-}; nutrients including sugars, fatty acids, and amino acids; and waste products, particularly carbon dioxide (CO_2), which must leave the cell.

The membrane's lipid bilayer structure provides the first level of control. The phospholipids are tightly packed together, and the membrane has a hydrophobic interior. This structure causes the membrane to be selectively permeable. A membrane that has selective permeability allows only substances meeting certain criteria to pass through it unaided. In the case of the cell membrane, only relatively small, nonpolar materials can move through the lipid bilayer (remember, the lipid tails of the membrane are nonpolar). Some examples of these are other lipids, oxygen and carbon dioxide gases, and alcohol. However, water-soluble materials—like glucose, amino acids, and electrolytes—need some assistance to cross the membrane because they are repelled by the hydrophobic tails of the phospholipid bilayer.

All substances that move through the membrane do so by one of two general methods, which are categorized based on whether or not energy is required. Passive transport is the movement of substances across the membrane without the expenditure of cellular energy. In contrast, active transport is the movement of substances across the membrane using energy from adenosine triphosphate (ATP).

2.2.1 Passive Transport

In order to understand how substances move passively across a cell membrane, it is necessary to understand concentration gradients and diffusion. A concentration gradient is the difference in concentration of a substance across a space. Molecules (or ions) will spread/diffuse from where they are more concentrated to where they are less concentrated until they are equally distributed in that space. (When molecules move in this way, they are said to move down their concentration gradient.) Diffusion is the movement of particles from an area of higher concentration to an area of lower concentration. A couple of common examples will help to illustrate this concept. Imagine being inside a closed bathroom. If a bottle of perfume were sprayed, the scent molecules would naturally diffuse from the spot where they left the bottle to all corners of the bathroom, and this diffusion would go on until no more concentration gradient remains. Another example is a spoonful of sugar placed in a cup of tea. Eventually the sugar will diffuse throughout the tea until no concentration gradient remains. In both cases, if the room is warmer or the tea hotter, diffusion occurs even faster as the molecules are bumping into each other and spreading out faster than at cooler temperatures. Having an internal body temperature around 98.6° F thus also aids in diffusion of particles within the body.

Fig.1: diffusion of particles

Whenever a substance exists in greater concentration on one side of a semipermeable membrane, such as the cell membranes, any substance that can move down its concentration gradient across the membrane will do so. Consider substances that can easily diffuse through the lipid bilayer of the cell membrane, such as the gases oxygen (O2) and CO2. O2 generally diffuses into cells because it is more concentrated outside of them, and CO2 typically diffuses out of cells because it is more concentrated inside of them. Neither of these examples requires any energy on the part of the cell, and therefore they use passive transport to move across the membrane.

Before moving on, you need to review the gases that can diffuse across a cell membrane. Because cells rapidly use up oxygen during metabolism, there is typically a lower concentration of O_2 inside the cell than outside. As a result, oxygen will diffuse from the interstitial fluid directly through the lipid bilayer of the membrane and into the cytoplasm within the cell. On the other hand, because cells produce CO_2 as a byproduct of metabolism, CO_2 concentrations rise within the cytoplasm; therefore, CO_2 will move from the cell through the lipid bilayer and into the interstitial fluid, where its concentration is lower. This mechanism of molecules moving across a cell membrane from the side where they are more concentrated to the side where they are less concentrated is a form of passive transport called simple diffusion.

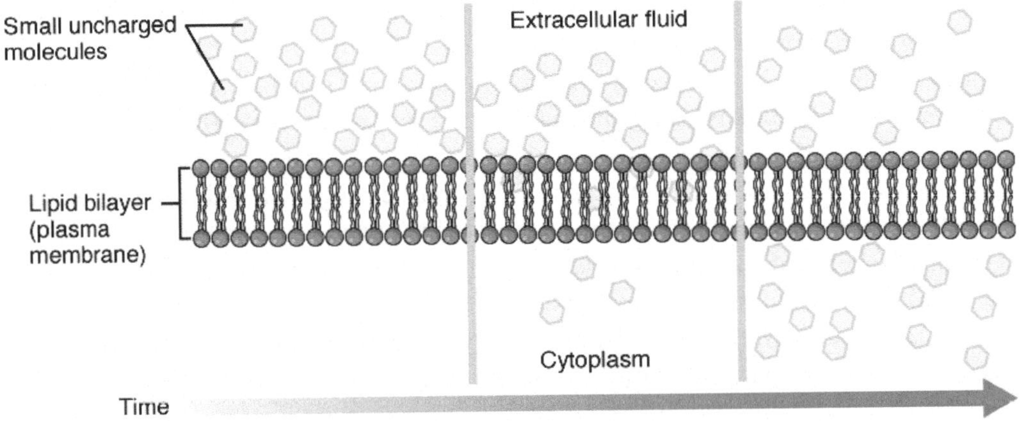

Fig.1: Simple Diffusion across Cell Membrane

Large polar or ionic molecules, which are hydrophilic, cannot easily cross the phospholipid bilayer. Very small polar molecules, such as water, can cross via simple diffusion due to their small size. Charged atoms or molecules of any size cannot cross the cell membrane via simple diffusion as the charges are repelled by the hydrophobic tails in the interior of the phospholipid bilayer. Solutes dissolved in water on either side of the cell membrane will tend to diffuse down their concentration gradients, but because most substances cannot pass freely through the lipid bilayer of the cell membrane, their movement is restricted to protein channels and specialized transport mechanisms in the membrane.

Facilitated diffusion is the diffusion process used for those substances that cannot cross the lipid bilayer due to their size, charge, and/or polarity. A common example of facilitated diffusion is the movement of glucose into the cell, where it is used to make ATP. Although glucose can be more concentrated outside of a cell, it cannot cross the lipid bilayer via simple diffusion because it is both large and polar. To resolve this, a specialized carrier protein called the glucose transporter will transfer glucose molecules into the cell to facilitate its inward diffusion.

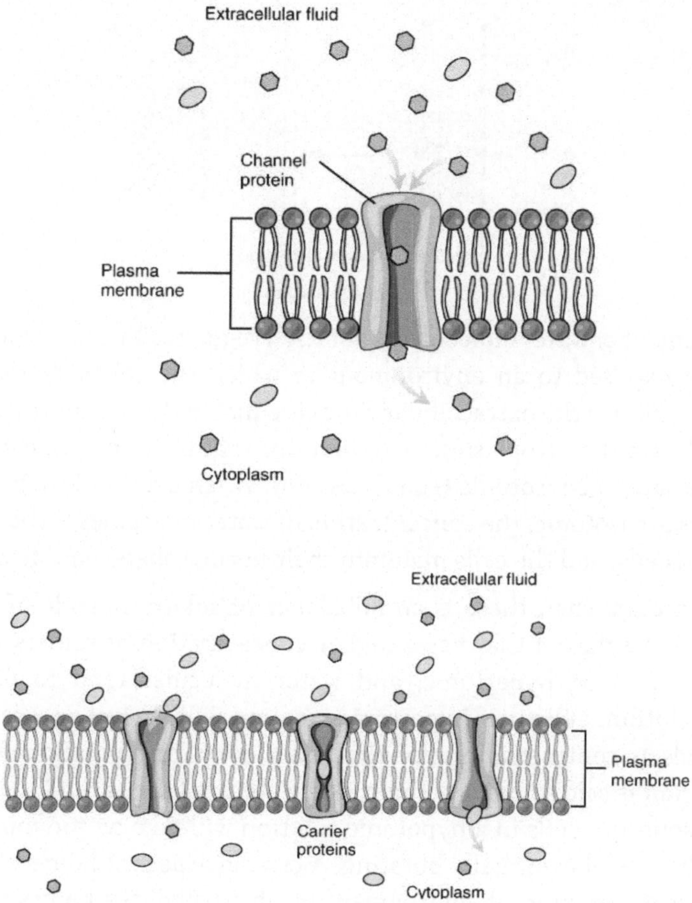

Fig. Facilitated Diffusion

As an example, even though sodium ions (Na+) are highly concentrated outside of cells, these electrolytes are charged and cannot pass through the nonpolar lipid bilayer of the membrane. Their diffusion is facilitated by membrane proteins that form sodium channels (or "pores"), so that Na+ions can move down their concentration gradient from outside the cells to inside the cells. There are many other solutes that must undergo facilitated diffusion to move into a cell, such as amino acids, or to move out of a cell, such as wastes. Because facilitated diffusion is a passive process, it does not require energy expenditure by the cell.

Water also can move freely across the cell membrane of all cells, either through protein channels or by slipping between the lipid tails of the membrane itself. Osmosis is the diffusion of water through a semipermeable membrane.

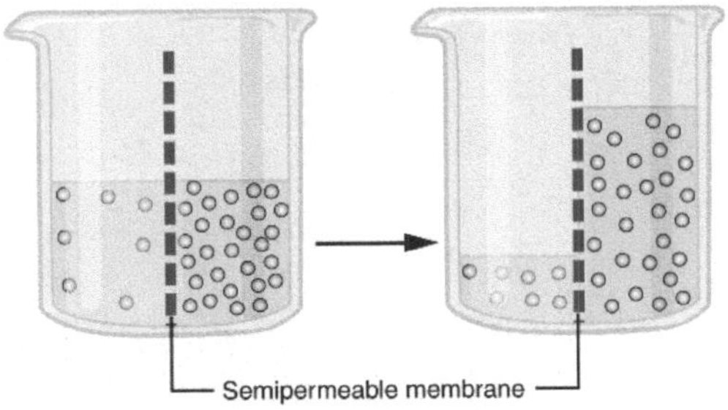

Fig.1: Osmosis

The movement of water molecules is not itself regulated by cells, so it is important that cells are exposed to an environment in which the concentration of solutes outside of the cells (in the extracellular fluid) is equal to the concentration of solutes inside the cells (in the cytoplasm). Two solutions that have the same concentration of solutes are said to be isotonic (equal tension). When cells and their extracellular environments are isotonic, the concentration of water molecules is the same outside and inside the cells, and the cells maintain their normal shape (and function).

Osmosis occurs when there is an imbalance of solutes outside of a cell versus inside the cell. A solution that has a higher concentration of solutes than another solution is said to be hypertonic, and water molecules tend to diffuse into a hypertonic solution. Cells in a hypertonic solution will shrivel as water leaves the cell via osmosis. In contrast, a solution that has a lower concentration of solutes than another solution is said to be hypotonic, and water molecules tend to diffuse out of a hypotonic solution. Cells in a hypotonic solution will take on too much water and swell, with the risk of eventually bursting. A critical aspect of homeostasis in living things is to create an internal environment in which all of the body's cells are in an isotonic solution. Various organ systems, particularly the kidneys, work to maintain this homeostasis.

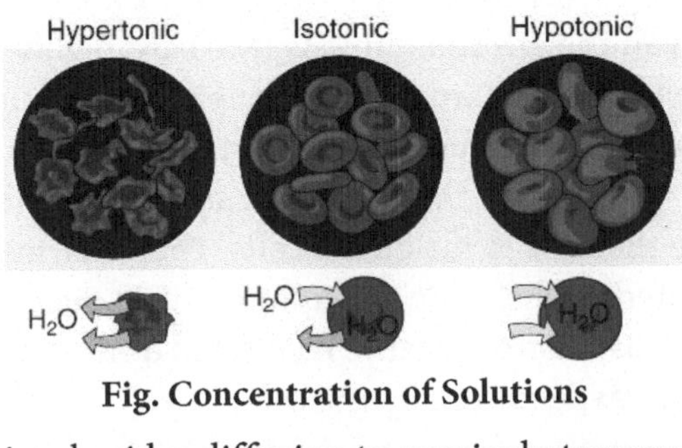

Fig. Concentration of Solutions

Another mechanism besides diffusion to passively transport materials between compartments is filtration. Unlike diffusion of a substance from where it is more concentrated to less concentrated, filtration uses a hydrostatic pressure gradient that pushes the fluid—and the solutes within it—from a higher pressure area to a lower pressure area. Filtration is an extremely important process in the body. For example, the circulatory system uses filtration to move plasma and substances across the endothelial lining of capillaries and into surrounding tissues, supplying cells with the nutrients. Filtration pressure in the kidneys provides the mechanism to remove wastes from the bloodstream.

2.2.2 Active Transport

For all of the transport methods described above, the cell expends no energy. Membrane proteins that aid in the passive transport of substances do so without the use of ATP. During active transport, ATP is required to move a substance across a membrane, often with the help of protein carriers, and usually against its concentration gradient.

One of the most common types of active transport involves proteins that serve as pumps. The word "pump" probably conjures up thoughts of using energy to pump up the tire of a bicycle or a basketball. Similarly, energy from ATP is required for these membrane proteins to transport substances—molecules or ions—across the membrane, usually against their concentration gradients (from an area of low concentration to an area of high concentration).

The sodium-potassium pump, which is also called Na^+/K^+ ATPase, transports sodium out of a cell while moving potassium into the cell. The Na^+/K^+ pump is an important ion pump found in the membranes of many types of cells. These pumps are particularly abundant in nerve cells, which are constantly pumping out sodium ions and pulling in potassium ions to maintain an electrical gradient across their cell membranes. An electrical gradient is a difference in electrical charge across a space. In the case of nerve cells, for example, the electrical gradient exists between the inside and outside of the cell, with the inside being negatively-charged (at around -70 mV) relative to the outside. The negative electrical gradient is maintained because each Na^+/K^+ pump moves three Na^+ions out of the cell and two K^+ ions into the cell for each ATP molecule that is used. This process is so important for nerve cells that it accounts for the majority of their ATP usage.

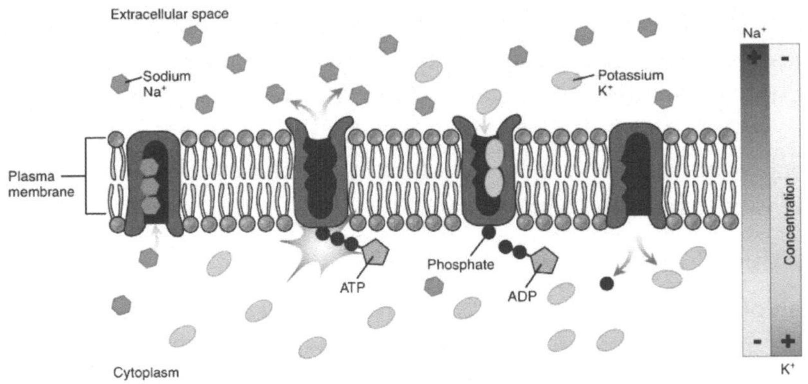

Fig. Sodium-Potassium Pump

Active transport pumps can also work together with other active or passive transport systems to move substances across the membrane. For example, the sodium-potassium pump maintains a high concentration of sodium ions outside of the cell. Therefore, if the cell needs sodium ions, all it has to do is open a passive sodium channel, as the concentration gradient of the sodium ions will drive them to diffuse into the cell. In this way, the action of an active transport pump (the sodium-potassium pump) powers the passive transport of sodium ions by creating a concentration gradient. When active transport powers the transport of another substance in this way, it is called secondary active transport.

Symporters are secondary active transporters that move two substances in the same direction. For example, the sodium-glucose symporter uses sodium ions to "pull" glucose molecules into the cell. Because cells store glucose for energy, glucose is typically at a higher concentration inside of the cell than outside. However, due to the action of the sodium-potassium pump, sodium ions will easily diffuse into the cell when the symporter is opened. The flood of sodium ions through the symporter provides the energy that allows glucose to move through the symporter and into the cell, against its concentration gradient.

Conversely, antiporters are secondary active transport systems that transport substances in opposite directions. For example, the sodium-hydrogen ion antiporter uses the energy from the inward flood of sodium ions to move hydrogen ions (H+) out of the cell. The sodium-hydrogen antiporter is used to maintain the pH of the cell's interior.

Other forms of active transport do not involve membrane carriers. Endocytosis (bringing "into the cell") is the process of a cell ingesting material by enveloping it in a portion of its cell membrane, and then pinching off that portion of membrane. Once pinched off, the portion of membrane and its contents becomes an independent, intracellular vesicle. A vesicle is a membranous sac—a spherical and hollow organelle bounded by a lipid bilayer membrane. Endocytosis often brings materials into the cell that must to be broken down or digested. Phagocytosis ("cell eating") is the endocytosis of large particles. Many immune cells engage in phagocytosis of invading pathogens. Like little Pac-men, their job is to patrol body tissues for unwanted matter, such as invading bacterial cells, phagocytize them, and digest them. In contrast to phagocytosis, pinocytosis ("cell drinking") brings fluid containing dissolved substances into a cell through membrane vesicles.

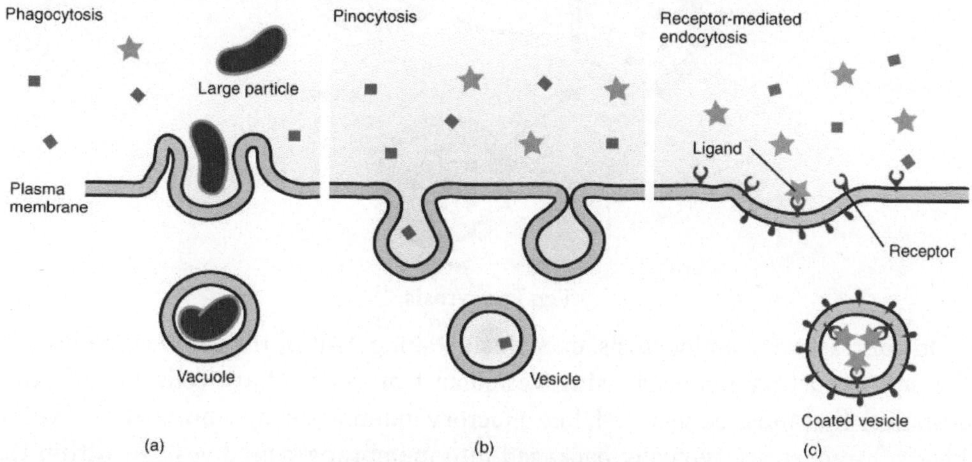

Fig. Forms of Endocytosis

Phagocytosis and pinocytosis take in large portions of extracellular material, and they are typically not highly selective in the substances they bring in. Cells regulate the endocytosis of specific substances via receptor-mediated endocytosis. Receptor-mediated endocytosis is endocytosis by a portion of the cell membrane that contains many receptors that are specific for a certain substance. Once the surface receptors have bound sufficient amounts of the specific substance (the receptor's ligand), the cell will endocytose the part of the cell membrane containing the receptor-ligand complexes. Iron, a required component of hemoglobin, is endocytosed by red blood cells in this way. Iron is bound to a protein called transferrin in the blood. Specific transferrin receptors on red blood cell surfaces bind the iron-transferrin molecules, and the cell endocytoses the receptor-ligand complexes.

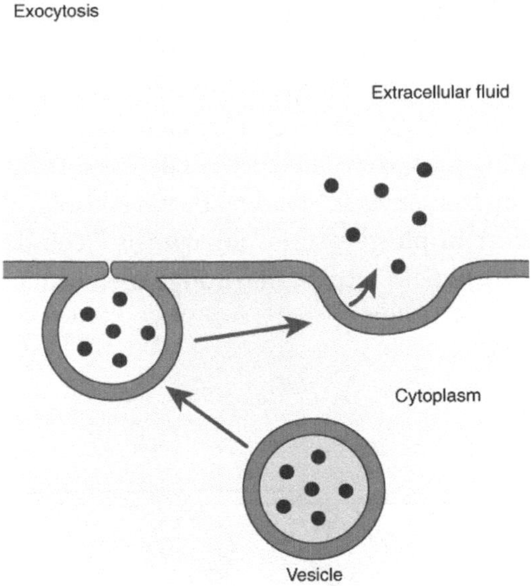

Fig. Exocytosis

In contrast with endocytosis, exocytosis (taking "out of the cell") is the process of a cell exporting material using vesicular transport. Many cells manufacture substances that must be secreted, like a factory manufacturing a product for export. These substances are typically packaged into membrane-bound vesicles within the cell. When the vesicle membrane fuses with the cell membrane, the vesicle releases it contents into the interstitial fluid. The vesicle membrane then becomes part of the cell membrane.

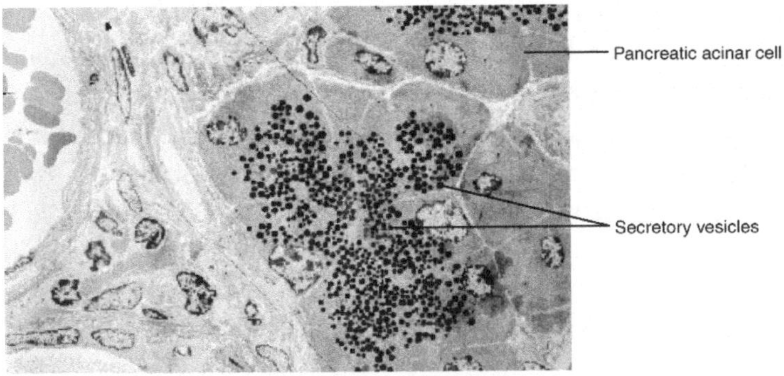

Fig. Pancreatic Cells

Cells of the stomach and pancreas produce and secrete digestive enzymes through exocytosis. Endocrine cells produce and secrete hormones that are sent throughout the body, and certain immune cells produce and secrete large amounts of histamine, a chemical important for immune responses.

GLOSSARY

active transport: form of transport across the cell membrane that requires input of cellular energy

amphipathic: describes a molecule that exhibits a difference in polarity between its two ends, resulting in a difference in water solubility

cell membrane: membrane surrounding all animal cells, composed of a lipid bilayer interspersed with various molecules; also known as plasma membrane

channel protein: membrane-spanning protein that has an inner pore which allows the passage of one or more substances

concentration gradient: difference in the concentration of a substance between two regions

diffusion: movement of a substance from an area of higher concentration to one of lower concentration

electrical gradient: difference in the electrical charge (potential) between two regions

endocytosis: import of material into the cell by formation of a membrane-bound vesicle

exocytosis: export of a substance out of a cell by formation of a membrane-bound vesicle

extracellular fluid: fluid exterior to cells; includes the interstitial fluid, blood plasma, and fluid found in other reservoirs in the body

facilitated diffusion: diffusion of a substance with the aid of a membrane protein

glycocalyx: coating of sugar molecules that surrounds the cell membrane

glycoprotein: protein that has one or more carbohydrates attached

hydrophilic: describes a substance or structure attracted to water

hydrophobic: describes a substance or structure repelled by water

hypertonic: describes a solution concentration that is higher than a reference concentration

hypotonic: describes a solution concentration that is lower than a reference concentration

integral protein: membrane-associated protein that spans the entire width of the lipid bilayer

interstitial fluid: fluid in the small spaces between cells not contained within blood vessels

intracellular fluid: fluid in the cytosol of cells

isotonic: describes a solution concentration that is the same as a reference concentration

ligand: molecule that binds with specificity to a specific receptor molecule

osmosis: diffusion of water molecules down their concentration gradient across a selectively permeable membrane

passive transport: form of transport across the cell membrane that does not require input of cellular energy

peripheral protein: membrane-associated protein that does not span the width of the lipid bilayer, but is attached peripherally to integral proteins, membrane lipids, or other components of the membrane

phagocytosis: endocytosis of large particles

pinocytosis: endocytosis of fluid

receptor: protein molecule that contains a binding site for another specific molecule (called a ligand)

receptor-mediated endocytosis: endocytosis of ligands attached to membrane-bound receptors

selective permeability: feature of any barrier that allows certain substances to cross but excludes others

sodium-potassium pump: (also, $Na+/K+$ ATP-ase) membrane-embedded protein pump that uses ATP to move $Na+$ out of a cell and $K+$ into the cell

vesicle: membrane-bound structure that contains materials within or outside of the cell

CHAPTER- 3

WATER

3.1 DEFINITION

Water has many unique properties that make it essential to all life. Most of water's unique properties are a result of the hydrogen bonding between water molecules. Water is an excellent solvent. When ionic compounds are placed into water, the ions dissociate or separate. Polar covalent compounds, because they too have charged poles, also dissolve in water. Nonpolar covalent compounds, however, do not dissolve in water. Thus polar covalent compounds are hydrophilic (water loving) while nonpolar covalent compounds are hydrophobic (water fearing).

Water has high cohesion. Individual water molecules tend to "stick" with other water molecules due to hydrogen bonding. This leads to two characteristics of water: a high surface tension and strong capillary action. Water's high surface tension is what allows some insects to run across the surface of the water. It is also the reason why belly-flops hurt so much. Water's strong capillary action is what allows the liquid level in a straw to be higher than that in the surrounding drink. Using hydrogen bonding, water molecules attract others up the sides of the straw. This effect is the more noticeable in thinner straws.

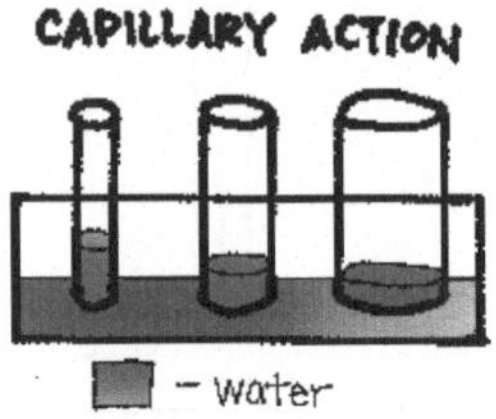

Fig Capillary action

Water has a high specific heat capacity. It takes a lot of energy to change the temperature of water. This characteristic makes water a good insulator and a good coolant. When you sweat, your body is using water as a coolant. The evaporating water removes heat with it. At night, oceans are a good insulator. The energy that the sun spent in heating the water all day is slowly released into the night.

Finally, water is the most dense at 4 degrees Celsius. Hydrogen bonding arranges water molecules into hollow "cells" when water freezes, making it less dense than liquid water. This characteristic of water is what allows fish in lakes and ponds to survive in the winter. When the water freezes, it becomes less dense and floats to the surface -- leaving the bottom of the lake or pond unfrozen.

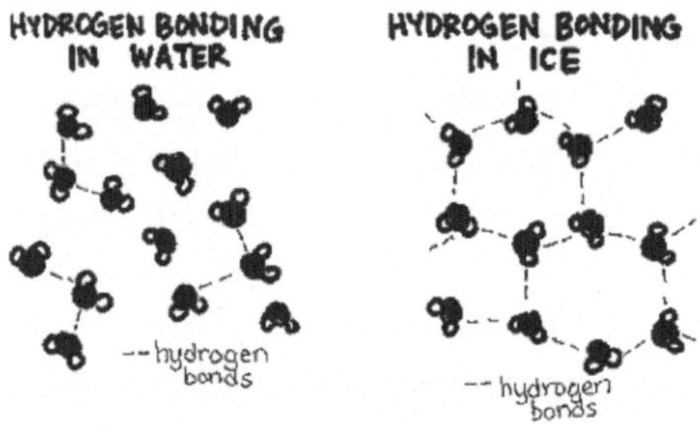

Fig Bonding

3.1.1 Weak interactions in aqueous systems

Water is a good solvent for polar solutes with which it forms hydrogen bonds. Non-polar compounds dissolve poorly in water. They can not form a hydrogen bond with the solvent. Numerous weak non covalent interactions influence the folding of macromolecules. The physical properties of aqueous solutions are strongly influenced by the concentration of solutes. Ex: The tendency for water to move across a semi permeable membrane is the osmotic pressure.

3.1.2 Ionization of water, weak acids and weak bases

Pure water is slightly ionized. pH is defined as the negative logarithm of the molar hydronium ion concentration. An acid that ionizes completely in water is a strong acid. An acid which is partially ionized in water is a weak acid. pk provides the measure of the ease of the dissociation of the proton. The lower the number the stronger is the acid.

3.1.3 Buffering against pH changes in biological systems

Buffers are aqueous systems that tend to resist in changes in pH when small amounts or acids or base are added. It consists of a weak acid and its conjugate base. As an example a mixture of equal concentrations of acetic acid and acetate ion is a buffer system. Every life form is extremely sensitive to slight pH changes. Human blood for example needs to remain within the pH range of 7.38 to 7.42. The relationship between pH and pk is described by Henderson – Hasselbach equation. It is used to calculate the pH of biological fluids. Two especially important biological buffers are the phosphate and bicarbonate systems. The property of a solution that depends on the concentration of a dissolved solute but not on its chemical identity is known as colligative properties. Substances that do not ionize in water are non electrolytes. Vapor pressure, boiling point, freezing point and osmosis are colligative properties.

3.2 WATER AS A REACTANT

Water is very often a direct participant in the chemical reactions of living cells. Formation of ATP from ADP and inorganic phosphate is an example of a condensation reaction in which the elements of water are eliminated. Two amino acid molecules are covalently joined through a peptide bond with a loss of water molecule.

3.2.1 Fitness of the aqueous environment for living organisms

The high specific heat of water is useful to cells and organisms because it allows water to act as a 'heat buffer', keeping the temperature of an organism relatively constant as the temperature of the surroundings fluctuates.

3.2.2 Features

Water is a good solvent for polar solutes with which it forms hydrogen bonds. Non-polar compounds dissolve poorly in water. They can not form a hydrogen bond with the solvent. Numerous weak non covalent interactions influence the folding of macromolecules. The physical properties of aqueous solutions are strongly influenced by the concentration of solutes. The equilibrium constant describes whether an acid or a base is strong or weak. A mixture of weak acid and its salt resists changes in pH caused by the addition of H^+ or OH^-. The mixture functions as a buffer. The relationship between pH and pk is described by Henderson-Hasselbach equation. It is used to calculate the pH of biological fluids. Water is very often a direct participant in the chemical reactions of living cells.

3.3 COMPOSITION OF WATER

Chemically pure water is composed of only two substances, hydrogen and oxygen. These materials in the free state exist only in a gaseous condition. Oxygen forms practically one-fifth of the volume of the atmosphere. Hydrogen does not exist very largely in a free state, though its occurrence in nature as hydrogen is sometimes observed. In nine pounds of pure water in round numbers eight pounds are oxygen and one pound hydrogen. If the composition of water is referred back to the gaseous state of its two components, it is found that two volumes of hydrogen combine with one volume of oxygen to form water. The union takes place on the application of a flame, with tremendous violence. The strongest containers of mixtures of hydrogen and oxygen in proportions to form water are ruptured with the greatest violence if the combination takes place.

The composition of water may be shown very beautifully and very easily by an experiment which produces its decomposition. If water containing an electrolyte such as diluted sulphuric acid is subjected to the action of the electric current hydrogen will be given off at the negative pole and oxygen at the positive pole of the electrical apparatus. If the negative pole be placed in one arm of a bent tube and the positive pole in the other, the gases may be collected separately. This is illustrated in the simple apparatus shown in the figure. The positive pole is marked + and is found to be giving off oxygen; the negative pole is marked - and is found to be giving off hydrogen. The gases rise in the tube and are collected. It will be seen that the volume of the hydrogen collected is just double that of the oxygen. This form of simple apparatus is called by chemists an eudiometer. If pure water with an electrolyte which cannot be decomposed is used the resulting gases will be pure.

Pure water is so poor a conductor of electricity that no decomposition will take place in the apparatus described because of the inability of the current to pass through the connecting arm filled with pure water between the positive and negative poles. The addition of a few drops of sulphuric acid enables the current to pass, while at the same time the acid is not in danger of being acted upon appreciably by the electric current itself. The result is that nothing is decomposed except pure water, so that pure hydrogen will be found in one arm of the apparatus and pure oxygen in the other. If the operation be conducted over an inverted tube in which both poles are in the same tube, the oxygen and hydrogen will escape together gradually forcing the liquid out at the bottom of the tube.

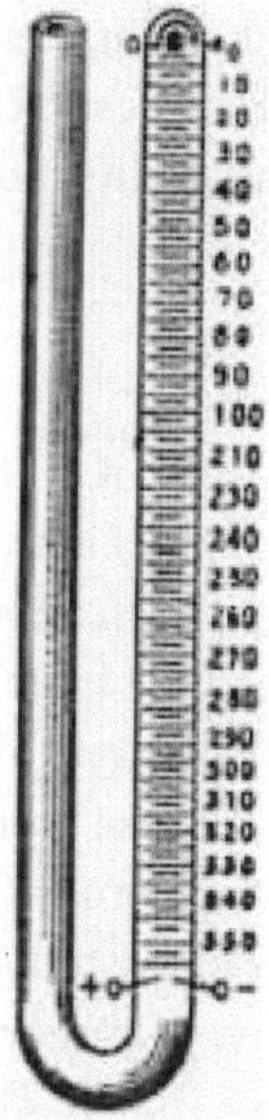

Fig. Eudiometer

3.4 RARE ELEMENTS PRESENT IN WATER

Attention has been called already to the fact that water is a great solvent. It acts with vigor upon many mineral substances in the soil and dissolves gases and other materials in the atmosphere. Water thus charged with other re-agents becomes more active on certain mineral substances. Thus the solvent power of water may be increased for some substances by dissolving other substances therein. This is notably illustrated in the case of limestone which is only slightly soluble in pure water but very soluble in water carrying carbon dioxid in solution. Such water passing through layers of limestone becomes heavily impregnated with dissolved carbonate of lime. On the water emerging from the spring or the well the carbon dioxid escapes and the excess of lime is precipitated. When such water is boiled in the tea kettle all of the carbon dioxid escapes and a part of the carbonate of lime is precipitated. In families where such water is used the tea kettle may often become incrusted with a layer of carbonate of lime. In addition to the ordinary substances found in water, those which are of more rare occurrence are frequently found. This is illustrated in the fact that some waters contain small quantities of lithium, in fact lithium seems to be a very widely distributed element and is found in many waters. As it will be seen further on, many waters claimed to be lithia waters are not so in fact, but have their lithium added after being taken from the source. There are genuine lithium waters. These waters are usually highly valued for medicinal purposes, though the value of the lithium a. therein as a remedial agent has been very much exaggerated. Some still rarer elements, or at least those less known, are often found in water. Among these may be mentioned argon and other rare constituents of the air. Certain waters also exhibit phases of radio-activity and have thus obtained a vogue as healing agents far beyond their merits.

3.5 BACTERIA IN WATER

The number of bacteria of all kinds in water varies enormously. After a long rain the last part of the precipitation may be almost free from impurities, including bacteria. The same is also true of water from snow collected on high mountain peaks, yet even in these cases absolute freedom from bacteria is not secured. The water that comes from great depths after filtration through rocks and beds of sands often has its bacterial content reduced to almost zero. The same is true of the waters of certain springs. When, however, water comes from any source near the surface of the soil its bacterial content is always likely to be high. The number of bacteria in the same water may vary from day to day. Warm and sultry days cause a great increase in the rapidity of bacterial growth, so that in changing from cold to warm weather suddenly the number of bacteria in any given water may rapidly change.

3.5.1 Pathogenic Bacteria

As has already been stated in the general discussion, there are many forms of diseases which are believed to be largely propagated through water. Among these the two which have been mentioned as most important are typhoid fever and cholera. The character of the organism which produces cholera is not so well understood as that which produces typhoid. The skilled bacteriologist will have no difficulty in recognizing the presence of typhoid germs and by isolating and cultivating them can reproduce them at will. Of course any water containing the typhoid germ is totally unfit for consumption. While, as has been intimated, such water may be used with impunity by some persons in vigorous health, it may at any moment seize upon one who consumes it, producing dangerous and even fatal results. The typhoid bacteria, when they do exist in water, are usually in very small numbers. It is therefore difficult to discover them. It is considered sufficient to discover other organisms which constantly accompany the typhoid germ, such as the colon bacillus and when these are found to assume that the typhoid germ may also be present and treat the water accordingly. There are many methods by which the typhoid germs enter the water. They may enter before consumption even when they are not present in the original source. When typhoid is epidemic all parts of the house and the vessels which may be in the house or near it may become infected. The water which may be brought into the house entirely free of typhoid germs, by being placed in vessels which are contaminated or polluted or otherwise exposed to infection, may become the vehicle to carry the germ in a dangerous way. The cholera bacillus is even more dangerous than that of typhoid. Fortunately it is rarely present in waters and therefore need not be feared except in cases of an epidemic. In all such cases no water should be regarded as above suspicion.

3.6 RAPIDITY OF MULTIPLICATION

The rapidity with which bacteria increase in number is sometimes phenomenal. Even in water where the bacterial food is not present in great abundance the increase is very rapid. In milk, which is a most favorable medium for rapidity of growth, the rate of increase is sometimes quite incredibly large. At ordinary room temperature the number of bacteria in any given amount of water increases at different rates. Franklin found, as quoted by George Newman (Bacteriology and the Public Health) that a water which contained in round numbers 1,000 organisms at the time of the collection of the sample had 6,000 at the end of 6 hours, 7,000 at the end of 24 hours, and 48,000 at the end of 48 hours. In another case where water was collected from a deep spring the number of organisms per cubic centimeter was found to be 7. In 24 hours the number had increased to 21. After three days the enormous number of 495,000 were found. This was at the usual temperature of the month of April. The kind of bacteria has also to be considered in this connection. Some varieties multiply much more rapidly than others.

3.6.1 Causes Which Produce Rapid Growth

There are two principal causes which produce very rapid growth of bacteria. The first of these is the presence in the medium of quantities of material upon which the organisms can feed; the second important factor is the temperature. All organisms, however, do not behave the same when at the same temperature. It has been found that organisms which grow in cold storage plants do not develop rapidly if at all at blood heat. On the other hand, organisms which develop normally at blood heat do not develop rapidly at a freezing temperature. It is believed that all these organisms, rapidly adapt themselves to their environments, so that after a few generations the organism which is normally grown at blood heat may develop at a low temperature, and vice versa.

3.6.2 Periodic Increase Or Decrease

All observers have noticed that the rate of increase or decrease in the bacterial count in water, as well as in other media, is irregular. There may be a rapid increase up to a very large number, then a sudden change showing no increase, or even a great falling off in the number of bacteria. This change may be due to various causes; for instance the amount of food which is at the disposal of the bacteria may become exhausted. In the second place, the bacteria by their decay and by the excretions of poisonous matters may render the medium germicidal and thus destroy their own life. Of course changes in temperature are likely also to increase or decrease the rate of multiplication.

3.7 WATER TREATMENT

Water treatment is a process of making water suitable for its application or returning its natural state. Thus, water treatment required before and after its application. The required treatment depends on the application. For example, treatment of grey water (from bath, dish and wash water) differs from the black water (from flush toilets). Composting toilet is not allowed in urban dwelling.

Water treatment involves science, engineering, business, and art. The treatment may include mechanical, physical, biological, and chemical methods. As with any technology, science is the foundation, and engineering makes sure that the technology works as designed. The appearance and application of water is an art.

In terms of business, RGF Environmental, Water Energy Technologies, Aquasana Store, Vitech, Recalyx Industrial SDN BHD and PACE Chemicals ltd are some of many companies that offer various processes for water treatment. Millipore, a Fisher Scientific partner, offers many lines of products to produce ultrapure water, using a combination of active charcoal membranes, and reverse osmosis filter. Internet sites of these companies offer useful information regarding water.

An environmental scientist or consultant matches the service provider, modify if necessary, with the requirement.

- ☆ Natural Water includes some discussion on hard and soft water. Softening hard water for boiler, cooler, and domestic application is discussed therein. These treatments prepare water so that it is suitable for the applications.
- ☆ Water Biology deals with water and biology. Drinking water is part of making water suitable for living. Thus, this link gives some considerations to drinking water problems.
- ☆ There are many different industry types, and waters from various sources are usually treated before and after their applications. Pre-application treatment and wastewater treatment offer a special opportunity or challenge. Only a general consideration will be given to some industrial processes.

General municipal and domestic wastewater treatment converts used water (waste) into environmentally acceptable water or even drinking water. Every urban centre requires such a facility.

3.7.1 General Wastewater Treatment

Water is a renewable resource. All water treatments involve the removal of solids, bacteria, algae, plants, inorganic compounds, and organic compounds. Removal of solids is usually done by filtration and sediment. Bacteria digestion is an important process to remove harmful pollutants. Converting used water into environmentally acceptable water or even drinking water is wastewater treatment.

Water in the Great Lakes Region is an organization dealing with the water resources. Ontario Clean Water Agency (OCWA) is a provincial Crown corporation in business to provide environmentally responsible and cost-efficient water and wastewater services. It currently operates more than 400 facilities for 200 municipalities. This web site provides information on water and water treatment.

Sewage Treatment

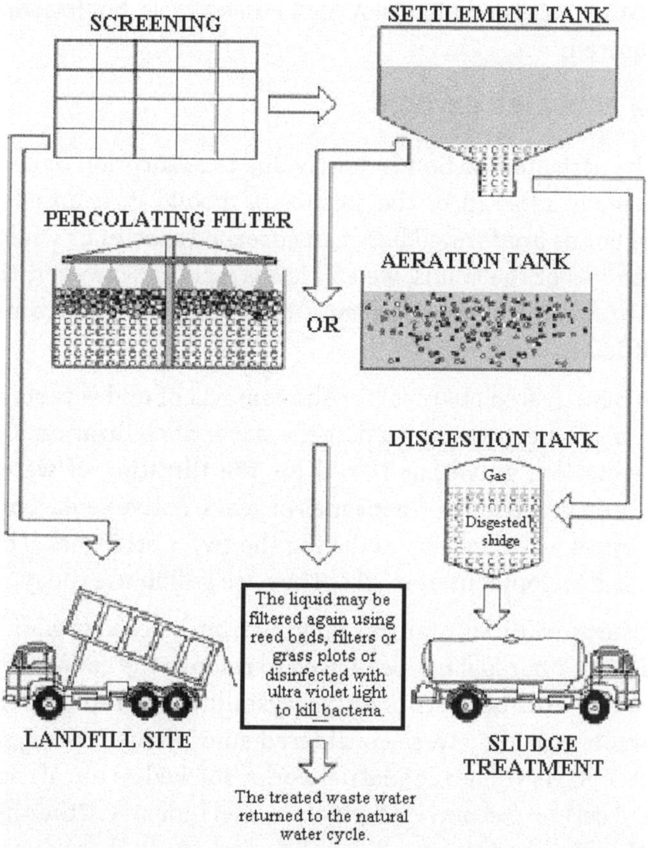

Sewage is SCREENED to remove large solid chunks, which are disposed in LANDFILL SITE. It flows over to the SETTLEMENT TANK to let the fine particles to settle. The settlement is called the activated SLUDGE. The supernatant is then PERCOLATING FILTERED and/or AERATED. The water can be filtered again, and then disinfected (chlorinated in most cases). When there is no other complication, the water is returned to nature back to the ecological cycle.

The SLUDGE removed from the settlement is composed of living biological material. A portion of it may be returned to the AERATION TANK, but the raw SLUDGE is digested by both microorganism. Anaerobic (without oxygen) and aerobic (with air) bacteria digestions are used. At the digestion stage, carbon dioxide, ammonia, and methane gases are evolved. Volume of the digested sludge is reduced, and it is acceptable as a fertilizer supplement in farming.

3.7.2 Wastewater Treatment

Although the sewage water may be discharged back to the ecological system after AERATED DIGESTION and PERCOLATING FILTRATION, but in some cases, further treatment is required.

Treatment by activated carbon

Treatment by activated carbon is mostly due to adsorption or absorption. When a chemical species is adhered to the surface of a solid, it is an adsorption. When partial chemical bonds are formed between adsorbed species or when the absorbate got into the channels of the solids, we call it absorption. However, these two terms are often used to mean the same, because to distinguish one from type from the other is very difficult.

Application of activated charcoal for the removal of undesirable order and taste in drinking water has been recognized at the dawn of civilization. Using bone char and charred vegetation, gravel, and sand for the filtration of water for domestic application has been practised for thousands of years. Active research and production of activated charcoal was accelerated during the two world wars. The use of poison gas prompted the development of masks. They are still in use today.

Charcoal absorbs many substances, ranging from colored organic particulates to inorganic metal ions. Charcoal has been used to remove the colour of raw sugar from various sources. Charcoal consists of microcrystallites of graphite. The particles are so small in charcoal that they were considered amorphous. The crystal structure of graphite consists of layers of hexagonal networks, stacked on top of each other. Today, making activated carbon is a new and widely varied industry. Other molecules attach themselves to the porous surface and dangling carbons in these microcrystallites.

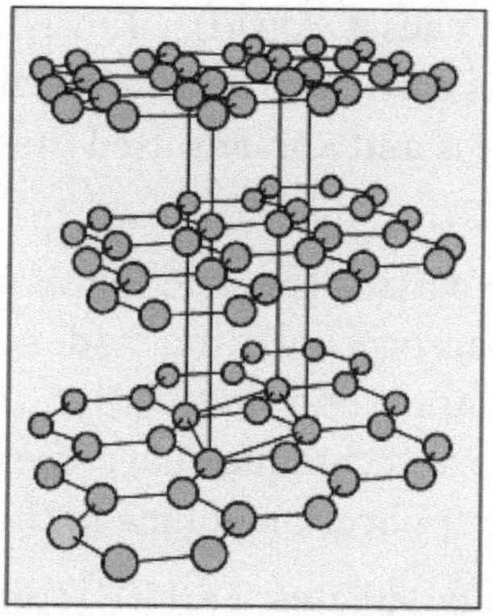

Fig crystal structure of graphite

Carbon containing substances are charred at less than 900 K to produce carbon in the manufacture of activated carbon. However, the carbon is activated at 1200 K using oxidizing agent to selectively oxidize portions of the char to produce pores in the material. Because of the special process to produce used, these materials with high surface to mass ratio, they are called activated carbon rather than activated charcoal. Factors affecting the absorption are particle size, surface area, pore structure, acidity (pH), temperature, and the nature of the material to be absorbed. Usually, adsorption (absorption) equilibria and rate of adsorption must be considered for effective applications.

Coagulation, flucculation and sedimentation

Natural and wastewater containing small particulates. They are suspended in water forming a colloid. These particles carry the same charges, and repulsion prevents them from combining into larger particulates to settle. Thus, some chemical and physical techniques are applied to help them settle. The phenomenon is known as coagulation. A well known method is the addition of electrolyte. Charged particulates combine with ions neutralizing the charges. The neutral particulates combine to form larger particles, and finally settle down.

Another method is to use high-molecular-weight material to attract or trap the particulates and settle down together. Such a process is called flocculation. Starch and multiply charged ions are often used.

Historically, dirty water is cleaned by treating with alum, Al2(SO4)3.12 H2O, and lime, Ca(OH)2. These electrolytes cause the pH of the water to change due to the following reactions:

Al2(SO_4) 3.12 H_2O, -> Al(aq)3++ 3 SO_4(aq)2- + 12 H_2O $SO_4(aq)^{2-}$+ H_2O -> HSO_4(aq)- + OH- (causing pH change) $Ca(OH)_2$ -> $Ca(aq)^{2+}$+ 2 OH-(causing pH change).

The slightly basic water causes $Al(OH)_3$, $Fe(OH)_3$ and $Fe(OH)_2$ to precipitate, bringing the small particulates with them and the water becomes clear. Some records have been found that Egyptians and Romans used these techniques as early as 2000 BC.

Suspension of iron oxide particulates and humic organic matter in water gives water the yellow muddy appearance. Both iron oxide particulates and organic matter can be removed from coagulation and flocculation. The description given here is oversimplified, and many more techniques have been applied in the treatment of water. Coagulation is a major application of lime in the treatment of wastewater.

Other salts such as iron sulfates $Fe_2(SO_4)_3$ and $FeSO_4$, chromium sulfate $Cr_2(SO_4)_3$, and some special polymers are also useful. Other ions such as sodium, chloride, calcium, magnesium, and potassium also affect the coagulation process. So do temperature, pH, and concentration.

Disposal of coagulation sludge is a concern, however.

Sedimentation let the water sit around to let the floculated or coagulated particles to settle out. It works best with relatively dense particles (e.g. silt and minerals), while flotation works better for lighter particles (e.g. algae, color). A settling tank should be big enough so that it takes a long time (ideally 4 hours +) to get through. Inlets and outlets are designed so the water moves slowly in the tank. Long and narrow channels are installed to let the water to snake its way through the tank. The settled particles, sludge, must occasionally be removed from the tanks. The water is next ready to be filtered. Sedimentation is used in pre-treatment and wastewater treatment.

Filtration

Filtration is the process of removing solids from a fluid by passing it through a porous medium. Coarse, medium, and fine porous media have been used depending on the requirement. The filter media are artificial membranes, nets, sand filter, and high technological filter systems. The choice of filters depends on the required filtering speed and the cleanness requirement. The flow required for filtration can be achieved using gravity or pressure. In pressure filtration, one side of the filter medium is at higher pressure than that of the other so that the filter plane has a pressure drop. Some portion of this filter type must be enclosed in a container.

The process of removing the clogged portion of the filter bed by reversing the flow through the bed and washing out the solid is called back washing. During this process, the solid must be removed out of the system, but otherwise the filters must be either replaced or taken out of service to be cleaned.

Aeration

Bringing air into intimate contact with water for the purpose of exchanging certain components between the two phases is called aeration. Oxygenation is one of the purposes of aeration. Others are removal of volatile organic substances, hydrogen sulfide, ammonia, and volatile organic compounds.

A gas or substance dissolved in water may further react with water. Such a reaction is called hydration. Ionic substance dissolve due to hydration, for example:

$$HCl\ (g) + x\ H_2O = H(H_2O)x+ + Cl(aq)-\ H_2S = H(aq)+ + HS(aq)-$$

These reactions are reversible, and aeration may also causes dehydration resulting in releasing the gas from water. Henry's law is applicable to this type of equilibrium for consideration. Methods of aeration are

- ☆ Diffused aeration - Air bubbles through water.
- ☆ Spray aeration - Water is sprayed through air.
- ☆ Multiple-tray aeration - Water flows through several trays to mix with air.
- ☆ Cascade aeration - Water flows downwards over many steps in the form of thin water falls.
- ☆ Air stripping - A combination of multiple tray and cascade technique plus random packed blocks causing water to mix thoroughly with air.

Reverse osmosis water filter system

In the following discussion, a dilute solution and a concentrated solution are considered. The dilute solution can be a clean water whereas the concentrated solution contains undesirable solute (electrolyte or others).

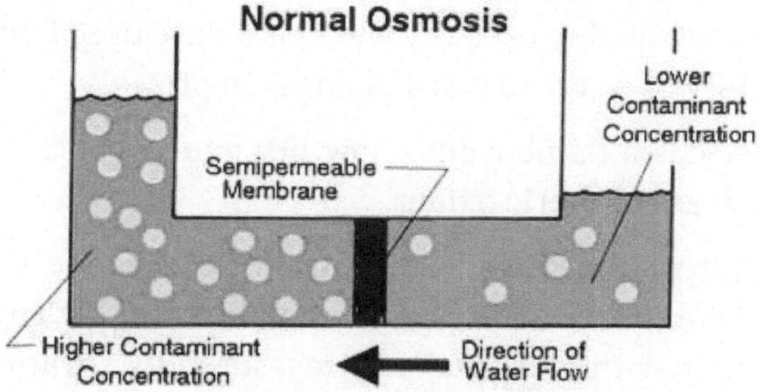

Fig Normal Osmosis

When a compartment containing a dilute solution is connected to another compartment containing a concentrated solution by a semipermeable membrane, water molecules move from the dilute solution to concentrated solution. This phenomenon is called osmosis. Pig bladders are natural semipermeable membranes.

As the water molecules migrate through the semipermeable membrane, water level in the solution will increase until the (osmotic) pressure prevents a net migration of water molecules in one direction. A pressure equivalent to the height difference is called the osmotic pressure.

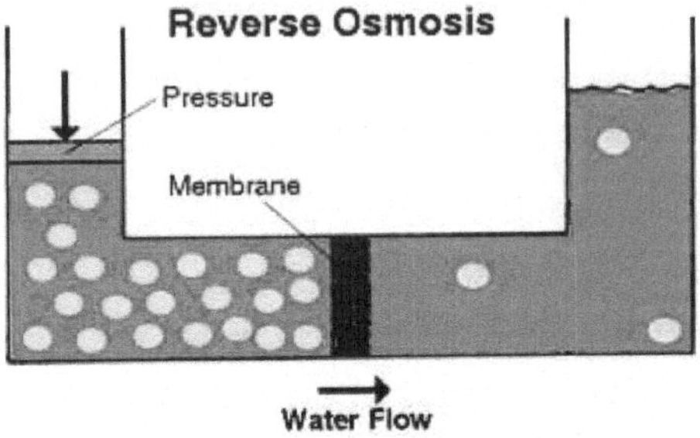

Fig.1: Reverse Osmosis

By applying pressure in the higher concentration solution, water molecules migrate from a high concentration solution to a low concentration solution. This method is called reverse osmosis water filter system. The concept of reverse osmosis is illustrated in the diagram here from PurePro.

GLOSSARY

Absolute: The micron rating of a filter. It indicates that any particle larger than a specific size will be trapped within the filter.

Absorption: When a solid takes up molecules into its structure.

Acid aerosol: Very small liquid or solid particles that are acidic and are small enough to become airborne.

Acidic water: Acid neutralizing capacity which is a measure of buffering capacity of water; the ability of water to resist changes in pH.

Acid rain: Rain that has a flamboyantly low pH, due to contact with atmospheric pollutants such as sulphuric oxides.

Acidity: The quantitative capacity of water to neutralize a base, expressed in ppm or mg/L calcium carbonate equivalent. The number of hydrogen atoms that are present determines this. It is usually measured by titration with a standard solution of sodium hydroxide.

Activated coal: This is the most commonly used adsorption medium, produced by heating carbonaceous substances or cellulose bases in the absence of air. It has a very porous structure and is commonly used to remove organic matter and dissolved gases from water. Its appearance is similar to coal or peat.

Available in granular, powder or block form; in powder form it has the highest adsorption capacity.

Activated sludge: Oxygen dependent biological process that serves to convert soluble organic matter to solid biomass, that is removable by gravity or filtration.

Active groups: Really fixed ions bolted on to the matrix of an ion exchanger. Each active group must always have a counter-ion of opposite charge near itself.

Adsorption: Separation of liquids, gases, colloids or suspended matter from a medium by adherence to the surface or pores of a solid.

Advanced oxidation process: One of several combination oxidation processes. Advanced chemical oxidation processes use (chemical) oxidants to reduce COD/BOD levels, and to remove both organic and oxidisable inorganic components. The processes can completely oxidise organic materials to carbon dioxide and water, although it is often not necessary to operate the processes to this level of treatment.

Advanced water treatment: The level of water treatment that requires an 85-percent reduction in pollutant concentration, also known as tertiary treatment.

Advanced Wastewater Treatment: Any treatment of sewage water that includes the removal of nutrients such as phosphorus and nitrogen and a high percentage of suspended solids.

Aerated lagoon: A water treatment pond that speeds up biological decomposition of organic waste by stimulating the growth and activity of bacteria, which are responsible for the degradation.

Aeration: Technique that is used with water treatment that demands oxygen supply, commonly known as aerobic biological water purification. Either water is brought into contact with water droplets by spraying or air is brought into contact with water by means of aeration facilities. Air is pressed through a body of water by bubbling and the water is supplied with oxygen.

Aeration tank: A tank that is used to inject air into water.

Aerobic: A process that takes place in the presence of oxygen, such as the digestion of organic matter by bacteria in an oxidation pond.

Aerosol: Very small liquid or solid particles dispersed in air.

Affinity: The keenness with which an ion exchanger takes up and holds on to a counter-ion. Affinities are very much affected by the concentration of the electrolyte surrounding the ion exchanger.

Agglomeration: A process of bringing smaller particles together to form a larger mass.

Aggressive water: Water that is soft and acidic and can corrode plumbing, pipes and appliances.

Algae: Single or multi-celled organisms that are commonly found in surface water, such as duckweed. They produce their own food through photosynthesis. The algae population is divided up into green algae and blue algae, of which the blue algae are very damageable to human health. Excessive algae growth may cause the water to have undesirable odours or tastes. Decay of algae diminishes oxygen supplies in the water.

Algal blooms: Periods of enlarged algal growths that affect water quality. Algal blooms indicate potentially hazardous changes in the chemistry of water.

Aliquot: A measured portion of a sample taken for analysis. One or more aliquots make up a sample.

Alkalinity: Alkalinity means the buffering capacity of water; the capacity of the water to neutralize itself. It prevents the water pH levels from becoming too basic or acid. It also adds carbon to water. Alkalinity stabilizes water at pH levels around 7. However, when the acidity is high in water the alkalinity decreases, which can cause harmful conditions for aquatic life.

Alluvium: Sediments deposited by erosion processes, usually by streams.

Anaerobic: A process that takes place in the absence of oxygen, such as the digestion of organic matter by bacteria in a UASB-reactor.

Anion: A negatively charged ion that results from the dissociation of salts, acids or alkali's in solution.

Anode: A site in electrolysis where metal goes into solution as a cation leaving behind an equivalent of electrons to be transferred to an opposite electrode, called a cathode.

Aquatic: Growing in water, living in water, or frequenting water.

Aqueous: Something made up of water.

Aqueous solubility: The maximum concentration of a chemical that dissolves in a given amount of water.

Aquifer: A layer in the soil that is capable of transporting a significant volume of groundwater.

Aromatics: A type of hydrocarbon that contains a ring structure, such as benzene and toluene. They can be found for instance in gasoline.

Assimilation: The ability of water to purify itself of pollutants.

Assimilative Capacity: The capacity of natural water to receive wastewaters or toxic materials without negative effects and without damage to aquatic life or humans who consume the water.

Atom: The smallest unit of matter that is unique to a particular element. They are the ultimate building blocks for all matter.

Attenuation: The process of reduction of a compound's concentration over time. This can be through absorption, adsorption, degradation, dilution or transformation.

Attrition: The action of one particle rubbing against the other in a filter media or ion exchange bed that can in time cause breakdown of the particles.

Available chlorine: A measure of the amount of chlorine available in chlorinated lime, hypochlorite compounds, and other materials.

Backflow: The flow of water in a medium in a direction opposite to normal flow. Flow is often returned into the system by backflow, if the wastewater in a purification system is severely contaminated.

Back Pressure: Pressure that can cause water to backflow into the water supply when a user's waste water system is at a higher pressure than the public system.

Back siphonage: Reverse seepage of water in a distribution system.

Backwashing: Reversing the flow of water back through the filter media to remove entrapped solids.

Bacteria: Microscopically small single-cell organisms, that reproduce by fission of spores.

Bacterial water contamination: The introduction of unwanted bacteria into a water body.

Base: An alkaline substance that has a pH that exceeds 7,5.

Bed Load: Sediment particles resting on or near the channel bottom that are pushed or rolled along by the flow of water.

Benthic zone: The lower region of a body of water including the bottom.

Bicarbonates: Salts containing the anion HCO3-. When acid is added, this ion breaks into H_2O and CO_2, and acts as a buffer.

Binder: Chemicals that hold short fibres together in a cartridge filter.

Bioaccumulantion: The increase in concentration of a substance in living organisms, as they take in contaminated air, water, or food, due to slow metabolization and excretion.

Biochemical Oxygen Demand (BOD): The amount of oxygen (measured in mg/L) that is required for the decomposition of organic matter by single-cell organisms, under test conditions. It is used to measure the amount of organic pollution in wastewater.

Biocide: A chemical that is toxic to microrganisms. Biocides are often used to eliminate bacteria and other single-cell organisms from water.

Biodegradable pollutants: Pollutants that are capable of decomposing under natural conditions.

Biofilm: Population of various microrganisms, trapped in a layer of slime and excretion products, attached to a surface.

Biological contaminants: Living organisms such as viruses, bacteria, fungi, and mammal and bird antigens that can cause harmful health effects to humans.

Biologically activated carbon: Activated carbon that supports active microbial growth, in order to aid in the degradation of organics that have been absorbed on its surface and in its pores.

Biological oxidation: Decomposition of complex organic materials by microrganisms through oxidation.

Biomonitoring: The use of living organisms to test the suitability of effluents for discharge into receiving waters and to test the quality of such waters downstream from the discharge.

Bioremediation: The biological treatment of wastewater and sludge, by inducing the breakdown of organics and hydrocarbons to carbon dioxide and water.

Biota: All living organisms in a region or ecosystem.

Biotransformation: Conversion of a substance into other compounds by organisms; including biodegradation.

Blackwater: Water that contains waste of humans, animals or food.

Blind spots: Any place on a filter medium where fluids cannot flow through.

Blinding: A build-up of particles in a filter medium, that prevents fluids from flowing through.

BOD5: The amount of dissolved oxygen consumed in five days by bacteria that perform biological degradation of organic matter.

Boiling point: The temperature at which the vapour pressure of a liquid equals the pressure of its surface. The liquid will than vaporize If the pressure of the liquid varies, the actual boiling point varies. For water the boiling point is 100 degrees Celsius.

Bottled water: Water that is sold in plastic containers for drinking water and/ or domestic use.

Brackish water: Water that is neither falls in the category of salt water, nor in the category of fresh water. It holds the middle between either one of the categories.

Breakpoint chlorination: Addition of chlorine to water until there is enough chlorine present for disinfection of water.

Breakthrough: Crack or break in a filter bed that allows the passage of floc or particulate matter through a filter.

Brine: Highly salty and heavily mineralised water, containing heavy metal and organic contaminants.

Buffer: A substance that reacts with hydrogen or hydroxyl ions in a solution, in order to prevent a change in pH.

Cake: Solid dewatered residue on a filter media after filtration.

Calcium hypo chlorite: A chemical that is widely used for water disinfection, for instance in swimming pools or water purification plants. It is especially useful because it is a stable dry powder and can be made into tablets.

Candle filter: A relatively coarse aperture filter, designed to retain a coat of filter medium on an extended surface.

Capillary action: Water that at some point rises higher than that portion of its surface, not in contact with the solid surface. This is due to adhesion, cohesion and surface tension where later touches a solid.

Cappilary membranes: Membranes about the thickness of a human hair, used for Reverse Osmosis, nanofiltration, ultrafiltration and microfiltrtion.

Capillary zone: Soil area above the water table where water can rise up slightly through the cohesive force of capillary action.

Carcinogen: Any dissolved pollutant that can induce cancer.

Cartridge filter: Disposable filter device that has a filter range of 0.1 micron to 100 microns.

Carbonates: Chemical compounds related to carbon dioxide.

Carbonate hardness: Hardness of water caused by carbonate and bicarbonate by-products of calcium and magnesium.

Catalyses: Chemical that increases the rate of a reaction but does not take a direct part in the reaction, so that it is still intact after the reaction has taken place.

Catch basin: A sedimentation area designed to remove pollutants from runoff before being discharged into a stream or pond.

Cathode: A site in electrolysis where cations in solution are neutralized by electrons that plate out on the surface or produce a secondary reaction with water.

Cation: A negatively charged ion, resulting from dissociation of molecules in solution.

Centrifugation: A separation process, which uses the action of centrifugal force to promote accelerated settling of particles in a solid-liquid mixture.

CFU: Colony Forming Units. This is a measure that indicates the number of microrganisms in water.

Check valve: A valve that allows water to stream in one direction and will then close to prevent development of a back-flow.

Chelating agents: Organic compounds that have the ability to draw ion from their water solutions into soluble complexes.

Chemical Oxygen Demand (COD): The amount of oxygen (measured in mg/L) that is consumed in the oxidation of organic and oxidasable inorganic matter, under test conditions. It is used to measure the total amount of organic and inorganic pollution in wastewater. Contrary to BOD, with COD practically all compounds are fully oxidized.

Chemical pollution: Introduction of chemical contaminants into a water body.

Chemical weathering: Dissolving of rock by exposure to rainwater, surface water, oxygen, and other gases in the atmosphere, and compounds secreted by organisms.

Chloramines: A chemical complex that consists of chlorine and ammonia. It serves as a water disinfectant in public water supplies in place of chlorine because chlorine can combine with organics to form dangerous reaction products. In which forms chloramines exist depends on the physical/ chemical properties of the water source.

Chlorinated hydrocarbons: Hydrocarbons that contain chlorine. These include a class of persistent insecticides that accumulate in the aquatic food chain. Among them are DDT, aldrin, dieldrin, heptachlor, chlordane, lindane, endrin, Mirex, hexachloride, and toxaphene.

Chlorinated solvent: An organic solvent containing chlorine atoms that is often used as aerosol spray container, in highway paint, and dry cleaning fluids.

Chlorination: A water purification process in which chlorine is added to water for disinfection, for the control of present microrganisms. It is also used in the oxidation of compound impurities in water.

Chlorine-contact chamber: The part of a water treatment plant where effluent is disinfected by chlorine.

Clarity: The clearness of a liquid.

Coagulation: Destabilisation of colloid particles by addition of a reactive chemical, called a coagulant. This happens through neutralization of the charges.

Coalescence: Liquid particles in suspension that unite to create particles of a greater volume.

Coastal zone: Lands and waters near the coast, whose uses and ecology are affected by the sea.

Coliform bacteria: Bacteria that serve as indicators of pollution and pathogens when found in water. These are usually found in the intestinal tract of humans and other warm-blooded animals.

Coliform index: A rating of the purity of water based on a count of coliform bacteria.

Collector sewers: Pipes to collect and carry wastewater from individual sources to an interceptor sewer that will carry it to a treatment facility.

Colloids: Matter of very small particle size, in the range of 10-5 to 10-7 in diameter.

Combined sewer: A sewer system that carries both sewage and rain water runoff.

Composite sample: A series of water samples taken over a given period of time and weighted by flow rate.

Compounds: Two or more different elements held together in fixed proportions by attractive forces called chemical bonds.

Concentrate: The totality of different substances that are left behind in a filter medium after filtration.

Concentration: The amount of material dissolved in a unit of solution, expressed in mg/L.

Concentration process: The process of increasing the number of particles per unit volume of a solution, usually by evaporating the liquid.

Condensate: Water obtained by condensation of water vapour.

Condensation: The change of state from a gas to a liquid.

Conductivity: The amount of electricity the water can conduct. It is expressed in a chemical magnitude. Please use also our information about TDS and conductivity.

Conduit: A natural or artificial channel through which fluids may be transported.

Consumptive water use: Water removed from available supplies without return to a water resources system; water used in manufacturing, agriculture, and food preparation.

Contact time: The length of time a substance is in contact with a liquid, before it is removed by filtration or the occurrence of a chemical change.

Contaminant: Any foreign component in a substance, for example in water.

Conventional sewer systems: Systems that were traditionally used to collect municipal wastewater in gravity sewers and convey it to a central primary or secondary treatment plant, before discharge on receiving surface waters.

Conveyance loss: Water loss in pipes and channels by leakage or evaporation.

Cooling tower: Large tower used to transfer the heat in cooling water from a power or industrial plant to the atmosphere either by direct evaporation or by convection and conduction.

Corrosivity: Ability of water to dissolve or break down certain substances, particularly metals.

Cross flow filtration: A process that uses opposite flows across a membrane surface to minimize particle build-up.

Cryptosporidium: A microorganism in water that causes gastrointestinal illness in humans. It is commonly found in untreated surface water and can be removed by filtration. It is resistant to disinfectants such as chlorine.

Cultural eutrophication: Decline of the oxygen rate in water, which has serious consequences for aquatic life, caused by humans.

Current: The portion of a stream or body of water, which is moving much faster than the rest of the water. The progress of the water is principally concentrated in the current.

Cycle: The length of time a filter can be used before it needs cleaning, usually including cleaning time.

Dealkalinisation: Any process that serves to reduce the alkalinity of water.

Decarbonation: The process of removing carbon dioxide from water, using contact towers or air scrubbers.

Decant: To draw off the upper layer of liquid after the heaviest material (a solid or another liquid) has settled.

Decomposition: The break down of organic matter by bacteria and fungi, to change the chemical structure and physical appearance of matter.

Defluoridation: The removal of fluoride from drinking water to prevent teeth damage.

De-foaming agents: Chemicals that are added to wastewater discharges to prevent the water from foaming when it is discharged into a receiving water body.

Degasification: The process of removing dissolved gasses from water, using vacuum or heat.

Deionisation: Process that serves to remove all ionised substances from a solution. Most commonly is the exchange process where cations and anions are removed independently of each other.

Demineralisation: Processes to remove minerals from water, usually the term is restricted to ion exchange processes.

Demiwater: Demineralised water. Water that is treated to be contaminant-, mineral- and salt free.

Denitrification: Removal of nitrate and nitrate product from water to produce a quality that answeres common water standards.

Density: The weight of a certain amount of water. It is usually expressed in kilograms per cubic metre.

Depression storage: The storage of water in low areas, such as ponds, and wetlands.

Depth filtration: Treatment process in which the entire filter bed is used to trap insoluble and suspended particles in its voids as water flows through it.

Desalination: The removal of salt from seawater or brackish water to produce drinking water, using various techniques.

Desorption: The opposite of adsorption; the release of matter from the adsorption medium, usually to recover material.

Detention time: The actual time that a small amount of water is in a settling basin or flocculating basin. In storage reservoirs, it means the length of time water will be stored.

Detergent: A water-soluble cleansing agent, other than soap.

Dewater: The separation of water from sludge, to produce a solid cake.

Diffuser: A component of the ozone contacting system in an ozone generator that allows diffusion of an ozone containing gas.

Diffusion: The movement of gas molecules or aerosols into liquids, caused by a concentration gradient.

Digester: A closed tank for wastewater treatment, in which bacterial action is induced to break down organic matter.

Diluting water: Distilled water that has been stabilized, buffered, and aerated. It is often applied in the BOD tests.

Direct run-off: Water that flows from the ground surface directly into streams, rivers, and lakes.

Discharge: Flow of surface water in a stream or canal.

Disinfectants: Fluids or gasses to disinfect filters, pipelines, systems, etc.

Disinfection: The decontamination of fluids and surfaces. To disinfect a fluid or surface a variety of techniques are used, such as ozone disinfection. Often disinfection means eliminating the present microrganisms with a biocide.

Dissolve: The process during which solid particles mix molecule by molecule with a liquid and appear to become part of the liquid.

Dissolved air flotation (DAF): A procedure of induced flotation with very fine air bubbles or 'micro bubbles', of 40 to 70 microns.

Dissolved oxygen: The amount of oxygen dissolved in water at a certain time, expressed in ppm mg/L.

Dissolved solids: Solids material that totally dissolves in water and can be removed by means of filtration.

Distillation: Water treatment method where water is boiled to steam and condensed in a separate reservoir. Contaminants with higher boiling points than water do not vaporize and remain in the boiling flask.

Dredging: Cleaning, deepening, or widening of a waterway, using a machine (dredge) that removes materials by means of a scoop or a suction device.

Drought: Term applied to periods of less than average precipitation over a certain period of time.

Duplicates: Two separate samples with separate containers taken at the same time and at the same place.

Dystrophic lakes: Acidic bodies of water that contain many plants but few fish, due to the presence of great amounts of organic matter.

Effluent: The outlet or outflow of any system that deals with water flows, for an oxidation pond for biological water purification. It is the product water of the given system.

Ejector: A device used to inject a chemical solution into wastewater during water treatment.

Electrical charge: The charge on an ion, declared by its number of electrons. A Cl- ion is in fact a Cl atom which has acquired an electron, and a Ca++ ion is a Ca atom, which has lost two electrons.

Electrolyte: Substance that dissociates into ions when it dissolves in water.

lectrodialysis: A process that uses electrical currents, applied to permeable membranes, to remove minerals from water.

Electrolysis: Process where electrical energy will change in chemical energy. The process happens in an electrolyte, a watery solution or a salt melting which gives the ions a possibility to transfer between two electrodes. The electrolyte is the connection between the two electrodes, which are also connected to a direct current. If you apply an electrical current, the positive ions migrate to the cathode while the negative ions will migrate to the anode. At the electrodes, the cations will be reduced and the anions will be oxidated.

Electrons: Negatively charged building blocks of an atom that circle around the nucleus.

Elements: The distinctive building blocks of matter that make up every material substance.

Elutriation: Freeing sludge of its mother liquor by washing it with water.

Emulsifier: A chemical that helps suspending one liquid in another.

Emulsion: Dispersion of one liquid in another liquid, occurs when a liquid in insoluble.

End-of-pipe techniques: Techniques for water purification that serve the reduction pollutants after they have formed.

Enrichment: When the addition of nutrients, such as nitrogen and phosphorus, from sewage effluent or agricultural runoff to surface water, greatly increases algal growth.

Erosion: The wearing away of the land surface by wind, water, ice or other geological agents. Erosion occurs naturally from weather or runoff but is often intensified by human land use practices.

Eutrophication: Enrichment of water, which causes excessive growth of aquatic plants and increasing activity of anaerobic microrganisms. As a result the oxygen levels in the water quickly decline and the water chokes, making life impossible for aerobic water organisms.

Evaporation: The process of the passage of water from liquid to vapour.

Evapotranspiration: The loss of water from the soil through vaporizing, both by direct evaporation and by transpiration from plants.

Facultative bacteria: Bacteria that can live under aerobic or anaerobic conditions.

Fermentation: The conversion of organic matter to methane, carbon dioxide and other molecules by anaerobic bacteria.

Filter medium: The permeable material that separates solids from liquids passing through it.

Filtrate: A liquid that has passed through the filter medium.

Filtration: Separation of a solid and a liquid by using a porous substance that only lets the liquid pass through.

First draw: The water that comes out when a tap is first opened. It is likely that is has the highest level of lead contamination from weathering of pipelines.

Fission: Reproduction of microrganisms by means of cell division.

Floc: A flocculent mass that is formed in the accumulation of suspended particles. It can occur naturally, but is usually induced in order to be able to remove certain particles from wastewater.

Flocculation: The accumulation of destabilized particles and micro flakes, and subsequently the formation of sizeable flakes. One must ad another chemical called flocculent in order to facilitate the formation of flakes called flocs.

Floodplain: The flat or nearly flat land along a river or stream that is covered by water during a flood.

Flotation: A solids-liquid or liquid-liquid separation procedure, which is applied to particles of which the density is lower than that of the liquid they are in. there are three types: natural, aided and induces flotation.

Flow: The discharge rate of a resource, expressed in volume during a certain period of time.

Flow augmentation: The addition of water to meet flow needs.

Flux: The rate at which a Reverse Osmosis Membrane allows water to pass through it.

Fouling: The deposition of organic matter on the membrane surface, which causes inefficiencies.

Fragmentation: The subdivision of a solid in fragments. The fragments will then adhere to the nearest surface.

Freezing: The change of a liquid into a solid as temperature decreases. For water, the freezing point is 0 degrees Celsius.

Freshwater: Water containing less than 1 mg/l of dissolved solids of any type.

Gallon: A unit that is now almost entirely out of date. It is equivalent to 3.785 litres.

Giardia: A microrganism that is commonly found in untreated surface water and can be removed by filtration. It is resistant to disinfectants such as chlorine.

Granular activated carbon: The heating of carbon to encourage active sites to absorb pollutants.

Gray Water: Domestic wastewater composed of wash water from kitchen, bathroom, and laundry sinks and from tubs, and washers.

Groundwater: Water that can be found in the saturated zone of the soil; a zone that consists merely of water. It slowly moves from places with high elevation and pressure to places with low elevation and pressure, such as rivers and lakes.

Groundwater discharge: Ground water entering coastal waters, which has been contaminated by land-fill leachates, deep well injection of hazardous wastes and septic tanks.

Groundwater hydrology: The branch of hydrology that deals with the occurrence, movements, replenishment and depletion, properties and methods of investigation and utilisation of groundwater.

Gully: A deeply eroded channel created by the concentrated flow of water.

Half-life: The time required for a pollutant to lose one-half of its original concentraton.

Hard water: Water that contains a great number of positive ions. The hardness is determined by the number of calcium and magnesium atoms present. Soap usually dissolves badly in hard water.

Heat exchanger: A component that is utilized to remove heat from or ad heat to a liquid.

Heavy metals: Metals that have a density of 5.0 or higher and a high elemental weight. Most are toxic to humans, even in low concentrations.

Heavy water: Water in which all the hydrogen atoms have been replaced by deuterium.

Henry's Law: A way of calculating the solubility of a gas in a liquid, based on temperature and partial pressure, by means of constants.

Holding Pond: A pond or reservoir, usually made of earth, built to store polluted runoff.

Homeowner water system: A water system that supplies piped water to a single residence.

Humidification: The addition of water vapour to air.

Hydraulic conductivity: The rate at which water can move through a permeable medium.

Hydraulic gradient: In general, the direction of groundwater flow due to changes in the depth of the water table.

Hydrocarbon: Organic compounds that are built of carbon and hydrogen atoms and are often used in petroleum industries.

Hydroelectric power water use: The use of water in the generation of electricity at plants where the turbine generators are driven by falling water.

Hydrogen sulfide (H2S): A gas emitted during organic decomposition by a select group of bacteria, which strongly smells like rotten eggs.

Hydrogeology: The science of chemistry and movement of groundwater.

Hydrograph: A chart that measures the amount of water flowing past a point as a function of time.

Hydrolysis: The decomposition of organic compounds by interaction with water.

Hydrophilic: Having an affinity for water.

Hydrophobic: Having an aversion for water.

Hydrosphere: Region that includes all the earth's liquid water, frozen water, floating ice, frozen upper layer of soil, and the small amounts of water vapour in the atmosphere.

Hypoxic waters: Waters with dissolved oxygen concentrations of less than 2 mg/L, the level generally accepted as the minimum required for life and reproduction of aquatic organisms.

Immiscibility: The inability of two or more solids or liquids to readily dissolve into one another.

mpermeable: Not easily penetrated by water.

Impurities: Particles or other objects that cause water to be unclear.

Indicator: Any biological entity or process, or community whose characteristics show the presence of specific environmental conditions or pollutants.

Indicator organisms: Microrganisms, such as coliforms, whose presence is indicative for pollution or for the presence more harmful microrganisms.

Indicator tests: Tests for a specific contaminant, group of contaminants, or constituent which signals the presence of something else.

Indirect discharge: Introduction of pollutants from a non-domestic source into a publicly owned wastewater treatment system. Indirect dischargers can be commercial or industrial facilities whose wastes enter local sewers.

Infiltration: Penetration of water into a medium, for instance the soil.

Influent: The stream of water that enters any system or treatment unit.

Inhibitor: chemical that interferes with a chemical reaction, such as precipitation.

Injection: The introduction of a chemical or medium into the process water to alter its chemistry or filter specific compounds.

Ion: An atom in a solution that is charged, either positively (cations) or negatively (anions).

Ion exchange: The replacement of undesirable ions with a certain charge by desirable ions of the same charge in a solution, by an ion-permeable absorbent.

Inorganic chemicals: Chemical substances of mineral origin, not of basically carbon structure.

Irrigation: Applying water or wastewater to land areas to supply the water and nutrient needs of plants.

Kinetic Energy: Energy possessed by moving water.

Kinetic rate coefficient: A number that describes the rate at which a water constituent such as a biochemical oxygen demand or dissolved oxygen rises or falls.

Laboratory water: Purified water used in the laboratory as a basis to create solutions or making dilutions. It contains no interfering substances.

Lagoon: A shallow pond where sunlight, bacterial action, and oxygen work to purify wastewater.

Laminar flow: A flow in which rapid fluctuations are absent.

Large water system: A water system that services more than 50,000 customers.

Leachate: Water that contains solute substances, so that it contains certain substances in solution after percolation through a filter or soil.

Leaching: The process by which soluble constituents are dissolved and filtered through the soil by a percolating fluid.

Leakage: A species of ions in the feed of an ion exchanger present in the effluent.

Light absorption: The amount of light a certain amount of water can absorb over time.

Lime: Common water treatment chemical. Lime can be deposed on walls of showers and bathrooms, after lime has reacted with calcium to form limestone.

Liquid: A state of matter, neither gas nor solid, that flows and takes the shape of its container.

Maximum Contaminant Level (MCL): The maximum level of a contaminant allowed in water by federal law. Based on health effects and currently available treatment methods.

Mechanical aeration: Use of mechanical energy to inject air into water to cause a waste stream to absorb oxygen.

echanical flotation: A term used in the mineral industry to describe the use of dispersed air to produce bubbles that measure 0.2 to 2 mm in diameter.

Medium-size water system: A water system that serves 3,300 to 50,000 customers.

Melting: The change of a solid into a liquid.

Membrane: A thin barrier that allows some compounds or liquids to pass through, and troubles others. It is a semi-permeable skin of which the pass-through is determined by size or special nature of the particles. Membranes are commonly used to separate substances.

Mesotrophic: Reservoirs and lakes which contain moderate quantities of nutrients and are moderately productive in terms of aquatic animal and plant life.

Metabolise: Conversion of food, for instance soluble organic matter, to cellular matter and gaseous by-products through a biological process.

MFS: Micro Filtration System, it serves full automatic solid/ liquid separation.

Microbial growth: The multiplication of microrganisms such as bacteria, algae, diatoms, plankton, and fungi.

Micron: A unit to discribe a measure of length, equal to one millionth of a metre.

Microrganisms: Organisms that are so small that they can only be observed through a microscope, for instance bacteria, fungi or yeasts.

Mineral Water: Contains large amounts of dissolved minerals such as calcium, sodium, magnesium, and iron. Some tap waters contain as many or more minerals than some commercial mineral waters. There is no scientific evidence that either high or low mineral content water is beneficial to humans.

Miscibility: The ability of two liquids to mix.

Mist: Liquid particles measuring 40 to 500 micrometers, are formed by condensation of vapour. By comparison, fog particles are smaller than 40 micrometers.

Mixture: Various elements, compounds or both, that are mixed.

Molecules: Combinations of two or more atoms of the same or different elements held together by chemical bonds.

Neutralization: The addition of substances to neutralize water, so that it is neither acid, nor basic. Neutralization does not specifically mean a pH of 7.0, it just means the equivalent point of an acid-base reaction.

Neutrons: Uncharged building blocks of an atom that play a part in radio-activity. They can be found in the nucleus.

Nitrification: A biological process, during which nitrifying bacteria convert toxic ammonia to less harmful nitrate. It is commonly used to remove nitrogen substances from wastewater, but in lakes and ponds it occurs naturally.

Non-point sources: Diffuse water pollution sources without a specific point of origin. The pollutants are generally carried off the land by storm water. Common non-point sources are agriculture and atmospheric disposal.

Non-potable: Water that is unsafe or unpalatable to drink because it contains pollutants, contaminants, minerals or infective agents.

Nucleus: The center of an atom, that contains protons and neutrons and carries a positive charge.

Nutrient: Any substance that promotes growth with living organisms. The term is generally applied to nitrogen and phosphorus in wastewater, but is also applied to other essential and trace elements.

Organic matter: Substances of (dead) plant or animal matter, with a carbon-hydrogen structure.

Osmosis: Water molecules passing through membranes naturally, to the side with the highest concentration of dissolved impurities.

Overflow rate: One of the guidelines for design of the settling tanks and clarifiers in a treatment plant to determine if tanks and clarifiers are used enough.

Oxidation: A chemical reaction in which ions are transferring electrons, to increase positive valence.

Ozone: An unstable oxidizing agent, that consists of three oxygen atoms and can be found in the ozone layer in the atmosphere. It is produced by electrical discharge through oxygen or by specifically designed UV-lamps.

Ozone generator: A device that generates ozone by passing a voltage through a chamber that contains oxygen. It is often used as a disinfection system.

Parameter: A variable, measurable property whose value is a determinant of the characteristics of a system such as water. Temperature, pressure, and density are examples of parameters.

Partial pressure: That pressure of a gas in a liquid, which is in equilibrium with the solution. In a mixture of gases, the partial pressure of any one gas is the total pressure times the fraction of the gas in the mixture (by volume or number of molecules).

Particle size: The sizes of a particle, determined by the smallest dimension, for instance a diameter. It is usually expressed in micron measurements.

Pasteurisation: The elimination of microrganisms by heat applies for a certain period of time.

Pathogens: Disease-producing microrganisms.

Percent saturation: The amount of a substance that is dissolved in a solution compared to the amount that could be dissolved in it.

Percolating water: Water that passes through rocks or soil under the force of gravity.

Permeability: The ability of a medium to pass a fluid under pressure.

Persistence: Refers to the length of time a compound stays in the environment, once introduced.

pH: The value that determines if a substance is acid, neutral or basic, calculated from the number of hydrogen ions present. It is measured on a scale from 0 to 14, on which 7 means the substance is neutral. pH values below 7 indicate that a substance is acidic and pH values above 7 indicate that it is basic.

Phase: A state of matter. This can be solid, liquid or gaseous.

Photosynthesis: The process of conversion of water and carbon dioxide to carbohydrates. It takes place in the presence of chlorophyll and is activated by sunlight. During the process oxygen is released. Only plants and a limited number of microrganisms can perform photosynthesis.

Physical and chemical treatment: Processes generally used in wastewater treatment facilities. Physical processes are for instance filtration. Chemical treatment can be coagulation, chlorination, or ozon treatment.

Physical weathering: Breaking down of rock into bits and pieces by exposure to temperature and changes and the physical action of moving ice and water, growing roots, and human activities such as farming and construction.

Phytoplankton: Free-floating, mostly microscopic aquatic plants.

Pilot tests: The testing of a cleanup technology under actual site conditions in a laboratory in order to identify potential problems before implementation.

POE-treatment: Point-Of-Entry treatment. Total water treatment at the inlet to an entire building or facility.

Pore: An opening in a membrane or medium that allows water to pass through.

Polar substance: A substance that carries a positive or negative charge, for instance water.

Pollutant: A contaminant at a concentration high enough to endanger the life of organisms.

POP's: Persistent Organic Pollutants, complex compounds that are very persistent and difficultly biologically degradable.

Potable water: Water that is safe for drinking and cooking.

Potentiation: The ability of one chemical to increase the effect of another chemical.

Potentiometric surface: The surface to which water in an aquifer can rise by hydrostatic pressure.

Precipitate: An insoluble reaction product in an aqueous chemical reaction.

Precipitation process: The altering of dissolved compounds to insoluble or badly soluble compounds, in order to be able to remove the compounds by means of filtration.

Pre-treatment: Processes used to reduce or eliminate wastewater pollutants from before they are discharged.

Primary wastewater treatment: The removal of suspended, floating and precipitated solids from untreated wastewater.

Protons: Positively charged building blocks of an atom that are centered in the nucleus.

Protozoa: Large microrganisms, which consume bacteria.

Public water system: A system that provides piped water for human consumption to at least 15 service connections or regularly serves 25 individuals.

Putrefaction: Biological decomposition of organic matter; associated with anaerobic conditions.

Pyrogen: Substance that is produces by bacteria and it fairly stable. It causes fever in mammals.

Qualitative water assessment: Analyses of water used to discribe the visible or aestetic charcteristics of water.

Quantitative water assessment: Use of analyses of water properties and concentrations of compounds and contaminants in order to define water quality.

Quicksilver water: A solution of mercury nitrate used in gilding.

Raw sewage: Untreated wastewater and its contents.

Raw water: Intake water before any treatment or use.

Reaeration: Renewing air supplies in the lower layers of a reservoir in order to raise oxygen levels.

Recarbonization: Process in which carbon dioxide is bubbled into treatment water in order to lower the pH.

Receiving waters: A river, lake, ocean, stream or other watercourse into which wastewater or treated effluent is discharged.

Run-Off: The part of precipitation water that runs off the land into streams or other surfacewater.

Safe water: Water that does not contain harmful bacteria, toxic materials, or chemicals, and is considered safe for drinking.

Safe yield: The annual amount of water that can be taken from a source of supply over a period of years without depleting that source beyond its ability to be naturally refilled.

Salinity: The presence of soluble minerals in water.

Sand filtration: Sand filtration is a frequently used and very robust method to remove suspended solids from water. The filtration medium consists of a multiple layer of sand with a variety in size and specific gravity. Sand filters can be supplied in different sizes and materials both hand operated and fully automatically.

Saturation: The condition of a liquid when it has taken into solution the maximum possible quantity of a given substance.

Scale: The precipitate that forms on surfaces in contact with water as the result of a physical or chemical change.

Screening: Use of screens to remove coarse floating and suspended solids from sewage.

Secondary treatment: The removal or reduction of contaminants and BOD of effluent from primary wastewater treatment.

Sedimentation: Settling of solid particles in a liquid system due to gravity.

Sediments: Soil, sand, and minerals washed from land into water, usually after rain.

Semi-confined aquifer: An aquifer partially confined by soil layers of low permeability through which recharge and discharge can still occur.

Semipermeable: A medium that allows water to pass through, but rejects dissolved solids, so that it can be used to separate solids from water.

Separate sewer: A sewer system that carries only sanitary sewage; no storm-water runoff. When a sewer is constructed this way, wastewater treatment plants can be sized to treat sanitary wastes only and all of the water entering the plant receives complete treatment at all times.

Separation: The isolation of the various compounds in a mixture.

Septic tank: An underground storage tank for wastes from homes not connected to a sewer line. Waste goes directly from the home to the tank.

Settleable solids: Those suspended solids in wastewater that will settle over a certain period of time and are removed in that way.

Settling: The process of sinking of a substance sinking in water. This occurs when the substance does not dissolve in water and its density is larger than that of water.

Sewage: Waste fluid in a sewer system.

Sewage contamination: The introduction of untreated sewage into a water body.

Sewage sludge: Sludge produced in a public sewer.

Sewerage: The entire system of sewage collection, treatment, and disposal.

Sludge: A semi-solid residue, containing microoroganisms and their products, from any water treatment process.

Softening: The removal of calcium and magnesium from water to reduce hardness.

Soft water: Any water that does not contain large concentrations of the dissolved minerals calcium or magnesium.

Solidification: Removal of wastewater from a waste or changing it chemically to make it less permeable and susceptible to transport by water.

Solubility: The amount of mass of a compound that will dissolve in a unit volume of water.

Solute: Matter dissolved in a liquid, such as water.

Solvent: Substance (usually liquid) capable of dissolving one or more other substances.

Sparger: A device that introduces compressed air into a liquid.

Sparging: Injection of air below the water table to strip dissolved volatile organic compounds and to facilitate aerobic biodegradation of organic compounds.

Specific conductance: Method to estimate the dissolved solid content of a water supply by testing its conductivity.

Spring: Ground water seeping out of the earth where the water table exceeds the ground surface.

Sublimation: The transitions of water directly from the solid state to the gaseous state, without passing through the liquid state.

Surface tension: The elastic-like force in a body, especially a liquid, tending to minimize, or constrict, the area of the surface.

Surface water: All water naturally open to the atmosphere, concerning rivers, lakes, reservoirs, ponds, streams, impoundments, seas, estuaries and wetlands.

Suspended solids: Solid organic or inorganic particles that are held in suspension in a solution.

Synergism: The combined action of several chemicals, which produces a total effect greater than the effects of the chemicals separately.

Tertiary treatment: Advanced cleaning of wastewater that goes beyond the secondary or biological stage, removing nutrients such as phosphorus, nitrogen, and most BOD and suspended solids.

Thermal pollution: Discharge of heated water from industrial processes in receiving surface water, causing death or injury of aquatic organisms.

TDS: Total Dissolved Solids. The weight per unit volume of water of suspended solids in a filter media after filtration or evaporation. Please use also our information about TDS and conductivity.

TH: Total Hardness. The sum of calcium and magnesium hardness, expressed as a calcium carbonate equivalent.

TS: Total Solids. The weight of all present solids per unit volume of water. It is usually determined by evaporation. The total weight concerns both dissolved and suspended organic and inorganic matter.

Total solids: All the solids in wastewater or sewage water, including suspended solids and filterable solids.

Toxic water pollutants: Compounds that are not naturally found in water at the given concentrations and that cause death, disease, or birth defects in organisms that ingest or absorb them.

Transpiration: The process by which water vapour is released into the atmosphere after transpiring of living plants.

Treatment plant: A structure built to treat wastewater before discharging it into the environment.

CHAPTER-4

CARBOHYDRATES

4.1 INTRODUCTION

The main function of carbohydrates in animals is to provide an energy source. A few carbohydrates serve as structural units such as deoxyribose. This sugar can be found in DNA. When the energy is not being used by the animal, the carbohydrate is stored as glycogen. Carbohydrates are composed of carbon, oxygen, and hydrogen. In most carbohydrates, there is one carbon atom for each water molecule.

4.2 CATEGORY OF CARBOHYDRATES

Carbohydrates are categorized into three groups based on molecule size.

4.2.1 Monosaccharides

Monosaccharides are composed of three to seven carbons. Prefixes indicate the number of carbon atoms contained in the molecule. For example, a pentose would consist of five carbon atoms. One of the main energy suppliers to the body happens to be a monosaccharide called glucose, a hexose.

4.2.2 Disaccharides

A disaccharide is the result when two monosaccharides combine by dehydration synthesis. One water molecule also results as a byproduct. For example, the monosaccharides fructose and glucose combine in the chemical equation C6H12O6(glucose) + C6H12O6(fructose) ----> C12H22O11(sucrose) + H2O(water) to form sucrose and water. You may notice that glucose and fructose have the same chemical composition; however they differ in the relative positions of their carbon and oxygen atoms. In order to obtain monosaccharides from a disaccharide, all you need to do is add water. This is a process called hydrolysis.

4.2.3 Polysaccharides

Polysaccharides consist of not only three or four monosaccharides, but tens and hundreds in long chains. These monosaccharides were also joined by dehydration synthesis. The main polysaccharide that is in the human body is glycogen. Glycogen is stored in the liver and skeletal muscles and can be broken down into monosaccharides for use as energy. In order to achieve monosaccharide status, polysaccharides must also go through hydrolysis.

4.3 LIPIDS (H)

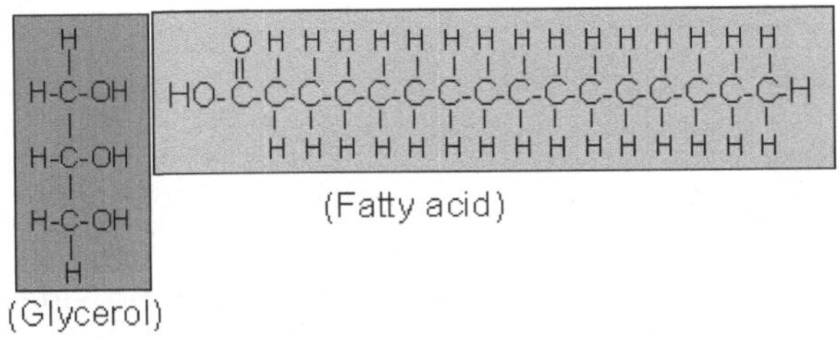

Fig Lipids

Lipids are very chemically diverse compunds which all share one common trait -- hydrophobia. They possess little or no affinity to water because of their molecular structure - mostly hydrocarbon regions with nonpolar bonds. There are three major categories of lipids:

4.3.1 Fats

large molecules composed of the two smaller molecules glycerol and fatty acids. When these two combine, the product is a triacylglycerol (or tryglyceride). The major function of fat is to supply a reserve of energy which is stored in adipose cells.

4.3.2 Phospholipids

Similar to fats, but contain two fatty acids and a phosphate group which attaches to the glycerol molecule. The phosphate group is hydrophilic while the hydrocarbons remain hydrophobic. This structure allows phospholipids to act as cell membranes inside the body.

4.4 STEROIDS

Consist of four interconnected carbon rings. Cholesterol, an important steroid, is also a part of cell membranes.

4.5 CARBOHYDRATES AS BIOMOLECULE

Carbohydrates are organic molecules with the general formula of CHO in 1:2:1 ratio. Although carbohydrates constitute only 1 to 2 percent of cell mass, they provide the raw fuel for cellular energy production. Carbohydrates are classified according to molecular size and solubility. In general, the smaller molecules are more soluble than the larger ones.

Monosaccharides include glucose, fructose, galactose, deoxyribose, and ribose.A single unit of sugar, monosaccharides are the smallest carbohydrates.Glucose, a 6-carbon sugar (hexose) is the sugar in our blood. Fructose, the sugar that sweetens fruit, and galactose, the sugar found in milk, have the same chemical formula as glucose and are therefore isomers of glucose. Isomers have the same composition but a different arrangement of their atoms and have different properties. Glucose can have a straight chain of carbon atoms or, more commonly, form a ring structure. Two other 5-carbon sugars or monosaccharides (called pentose) used in nucleic acid synthesis are deoxyribose and ribose.

Disaccharides include sucrose, lactose, and maltose.Two monosaccharides are joined together by dehydration synthesis to form a disaccharide molecule.Sucrose (table sugar)Glucose + fructose = sucrose + waterLactose (milk sugar)Glucose + galactose = lactose + waterMaltose (malt sugar)Glucose + glucose = maltose + water

Polysaccharides include starch, cellulose, and glycogen.These long, chain-like polymers make ideal storage products due to their insolubility.Starch is the storage molecule synthesized from glucose by plants.Cellulose, which is also synthesized by plants for cell wall construction, is indigestible because we lack enzymes for it. Cellulose provides fiber to promote peristalsis.Glycogen is the carbohydrate storage molecule found in muscle and liver cells. When blood sugar levels drop, liver cells hydrolyze glycogen and release glucose to the blood.

4.6 CLASSIFICATION

The activities on which the life of the body depends involve a continuous expenditure of energy and a constant exchange of material. Ultimately the body is dependent upon food for the fuel materials which supply energy and for both the substances which are transformed in, and eliminated from, the body, and those whose presence regulates and controls these transformations. The materials leaving the body are to be regarded not merely as wastes but as end products of an orderly and coordinated series of chemical reactions which occur in the body and by virtue of which its functions are performed. Thus the chief functions of food are: (1) to yield energy, (2) to build tissue, (3) to regulate body processes.

These functions involve reactions which are dependent upon the chemical composition and constitution of the food. Any food constituent which takes part in any of these functions may be regarded as having nutritive value.

Most of the nutrient material contained in food requires more or less change to bring it into the exact forms most useful in nutrition. These changes as a rule take place in the digestive tract and together constitute the process of digestion.

The changes which take place in the foodstuffs, after they have been absorbed from the digestive tract, are included under the general term "metabolism." Although the chemical changes and the energy transformations are of course inseparable, it has become customary to speak of the metabolism of matter and the metabolism of energy, and to regard the extent of the metabolism of any material substance as measured by the amount of its end products eliminated, and the extent of the energy metabolism as measured by the amount of heat, or of heat and external muscular work, which the body gives off.

The metabolism of matter and the metabolism of energy are normally supported by the food; but if no food is taken, they continue at the expense of the body substance. The expenditure of energy can never cease in the living body because it includes the work involved in carrying on the internal processes which are essential to life itself; and the expenditure of matter cannot cease because the energy for this necessary work is obtained by the breaking down of the organic compounds of the food or of the body substance into simpler compounds, many of which are of no further use to the body and must be eliminated. When the food supplies sufficient energy, the body substance is protected; when the food is insufficient, body substance is burned as fuel. In order, then, to consider intelligently the nutritive requirements of the body as regards the substances of which it is composed, it is necessary first to know whether the fuel requirements (the requirements of the energy metabolism) have been fully met.

The carbohydrates, fats, and proteins of the food all serve as fuel to yield the energy required for the activities of the body, and the proteins serve also as material for the maintenance or growth of body tissue. But of the fifteen chemical elements which are essential to the structure and functions of the body, simple proteins furnish only five. The remaining ten elements are largely constituents of the ash of the food and are known as ash constituents, inorganic foodstuffs, mineral matter, or salts.

Recent investigations have developed the fact that food of sufficient energy value and containing ample amounts of each of the chemical elements known to be essential to the body is not necessarily adequate to meet all the requirements of nutrition. Thus it appears that certain substances occurring in natural foods but not yet chemically identified are also to be included among the nutritive requirements of the body and therefore among the factors which determine the nutritive values of foods. At present these unidentified substances are referred to as "vitamines" or as "fat soluble A" and "water soluble B."

The essentials of a chemically adequate food supply may therefore be summarized as follows: (1) sufficient of the organic nutrients in digestible forms to yield the needed energy; (2) protein, sufficient in amount and appropriate in kind; (3) adequate amounts and proper proportions of the various ash constituents or inorganic foodstuffs; (4) sufficient of each of the two unidentified "vitamine" factors, the "fat soluble A" and the "water soluble B."

In attempting to give in the following pages a general view of the chemistry of food and nutrition it has seemed best to discuss first the chemical nature and nutritive functions of the substances which serve as sources of energy in nutrition, then the nutritive requirements in terms of energy, protein, the more prominent "inorganic" elements, and the "vitamines," and finally the bearing of these various factors offood value upon problems connected with the economic use of food.

Of the constituents of the ordinary mixed food of man the carbohydrates are usually the most abundant and the most economical sources of energy. They are also considered to be the first of the three great groups of foodstuffs to be formed by synthesis from simple inorganic substances in plants; "in the long run, all the energy of living matter comes from them." The synthesis of carbohydrates in nature is therefore a logical starting point for the study of the organic foodstuffs.

In the chlorophyll cells of the leaves of green plants the energy of the sun's rays brings about reaction between carbon dioxide and water which results in the liberation of oxygen and the formation of organic compounds. There is still doubt as to the exact mechanism of the process and no certainty that it is the same in all cases. It has, however, been quite generally found that the volume of oxygen liberated is equal to that of carbon dioxide consumed. The simplest possible representation of the reaction would be $CO2 + H20 \rightarrow CH20 + O2$ according to which the first product of the synthesis would be formaldehyde. There is considerable (though not conclusive) evidence that formaldehyde is thus formed and that it is rapidly built up into less reactive compounds. Whatever the steps in the process* there is normally an early production of carbohydrate. Usually the first product which can be demonstrated as accumulating in the plant as the result of the photosynthesis is a sugar (glucose or sucrose) or starch. Assuming glucose as a typical product and neglecting the intermediate stages, the photosynthesis of carbohydrate may be represented thus:

$$6\ CO^2 + 6\ H^20 \rightarrow C^6H^{12}O^6 + 6\ O^2$$

Glucose is the most familiar representative of a group of simple sugars (monosaccharides or monosaccharoses) which are in composition direct polymers of formaldehyde (CH_2O) and which are classified, according to the number of carbon atoms in the monosaccharide molecule, as trioses, pentoses, hexoses, etc.

According to Browne, a simple sugar or monosaccharide may be defined as an aldehyde alcohol or ketone alcohol of the aliphatic series, the molecule of which contains one carbonyl and one or more alcohol groups, one of the latter being always adjacent to the carbonyl group. All simple sugars contain, therefore, as a characteristic group upon the presence of which the chief chemical properties of the sugars depend.

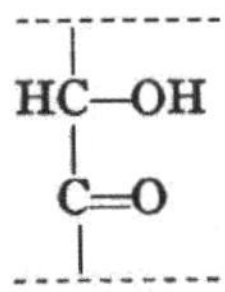

The simplest possible sugar according to this definition is glycolaldehyde, CH_2OH-CHO, which (in analogy with the nomenclature of the familiar sugars) may also be called glyco-lose. The structural formulae of glucose and fructose, the most familiar representatives of the aldehyde-alcohol (aldose) and ketone-alcohol (ketose) sugars, respectively, are as follows:

4.6.1 Glucose

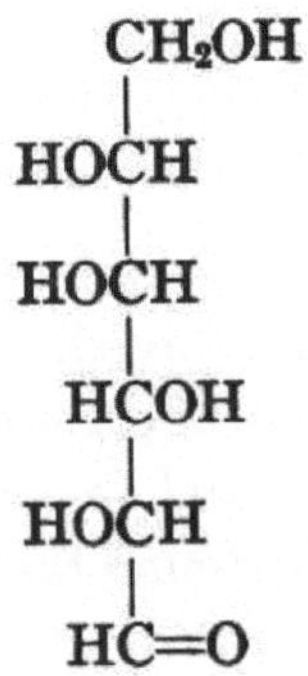

4.6.2 Fructose

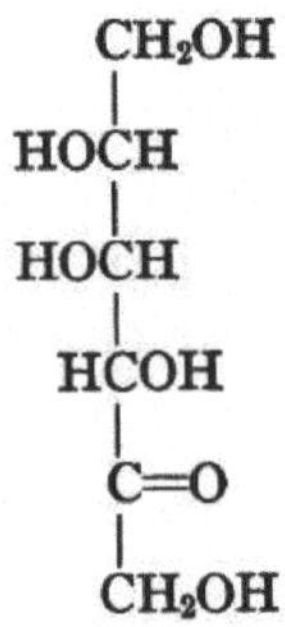

Since glucose gives aldehyde reactions but not so readily as the above structural formula would lead one to expect, it is believed that in ordinary solutions of glucose the substance exists partly in the condition indicated by the aldehyde formula and partly in a tautomeric form represented by the lactone or "oxygen bridge" formula. Following are the aldehyde and lactone formulae written without reference to the spatial relationships of the hydrogen and hydroxyl groups:

Aldehyde form:

CH2OH-CHOH-CHOH-CHOH-CHOH-CHO

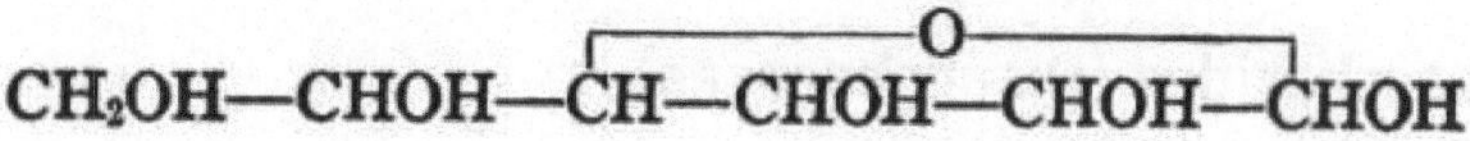

The name monosaccharide ("single sugar") implies that the monosaccharide molecule contains only one sugar radicle - that it cannot be split by hydrolysis into sugars of lower molecular weight. A substance like cane sugar which on hydrolysis splits to two molecules of simple sugar is called a disaccharide or disaccha-rose ("double sugar"). Trisaccharides and tetrasaccharides are also known. Substances which like starch are of high molecular weight and on complete hydrolysis yield many molecules of simple sugar are called polysaccharides * or polysaccharoses. The term "carbohydrates" covers all the simple sugars and all substances which can be converted into simple sugars by hydrolysis. The term "glucosides" is applied to substances which consist of combinations of carbohydrate radicles with radicles of other kinds and which therefore yield on hydrolysis both a simple sugar and one or more products of other than carbohydrate nature.

4.6.3 Monosaccharides

(Monosaccharoses) Dioses (C2H4O2) - Glycolose.

Moses (C3H6O3). Aldoses - Glycerose. Ketose - Dioxyacetone.

Tetroses (C4H8O4).

Aldoses - Erythrose,3 Threose.3 Ketose - Erythrulose.2

Pentoses (C6H10O5).

Aldoses - Arabinose,2 Xylose,2 Ribose,2 Lyxose.8 Ketoses - Araboketose3, Xyloketose (ketoxylose).2 [Methyl pentoses (C6H10O5) - Rhamnose,2 Fucose2].

Some writers use the term polysaccharides to include all carbohydrates other than monosaccharides. Mathews applies it to all carbohydrates more complex than the disaccharides.

Names of a few of the most important carbohydrates are printed in small capitals. Separate mention of the d, l, and dl forms of the various sugars is omitted, since in the study of food and nutrition we are practically concerned only with that one of the three forms which is found in or derived from natural products.

Hexoses (C6H12O6).

Aldoses - Glucose,1 Mannose,2 Galactose,2 Gulose,3 Idose,3Talose,3

Allose,2 Altrose.3 Ketoses - Fructose,1 Sorbose,2 Tagatose.3

Heptoses (C7H14O7). Aldose - Mannoheptose.1 Ketose - Sedoheptose.1

4.6.4 Disaccharides (Disaccharoses)

Dihexoses (Hexobioses) - (C12H22O11). Anhydride of glucose + fructose-Sucrose. Anhydrides of glucose + galactose - Lactose, Melibiose. Anhydrides of glucose + glucose- Maltose, Isomaltose, Trehalose, Turanose.

4.6.5 Trisaccharides (Trisaccharoses)

Trihezoses (C18H22O16). Anhydride of glucose + galactose +fructose - Raffinose. Anhydride of glucose + glucose + glucose - Melezitose. Anhydride of fructose + fructose + fructose - Secalose.

4.6.7 Tetrasaccharides (Tetrasaccharoses)

Tetrahexoses (C24H42O21). Anhydrides of 2 galactose + glucose + fructose - Stachyose, Lupeose.

4.6.8 Polysaccharides (Polysaccharoses)

Pentosans (chief constituents of gums and mucilages). Anhydrides of xylose - Xylans. Anhydrides of arabinose - Arabans.

Anhydrides of glucose-Starch, Cellulose, Glycogen, Dextrin (and other "dextrans"). Anhydrides of mannose - Mannans. Anhydrides of galactose- Galactans (pectins). Anhydrides of fructose - Inulin (and other "levulans").

- ☆ Occurs free in nature.
- ☆ Not yet found free in nature (or only in small amounts) but obtained by hydrolysis or fermentation of natural product.
- ☆ Known only (with certainty) as a laboratory product.

4.7 BIOCHEMICAL TESTS ON CARBOHYDRATES

4.7.1 Materials

- ☆ Benedict's solution (not "quantitative")

- ☆ Iodine solution (I/KI)
- ☆ Waterbaths set at about 100°C
- ☆ Boiling tubes Racks
- ☆ 1% glucose soln 1% fructose soln 1% sucrose soln other sugars?
- ☆ 1% starch sol solution X
- ☆ HCl, NaOH, litmus/universal indicator paper filter funnels, papers, flasks? measuring cylinders (10, 25, 50, 100 ml) plastic pipettes glass pipettes (1, 5, 10 ml?), pipette pumps

4.7.2 Objectives

Familiarisation with Benedict's test - for reducing sugars modification of Benedict's test for non-reducing sugars iodine test dilution techniques sensitivity of above tests

- ☆ Carry out Benedict's test on the named sugar solutions and starch as provided. Use fixed (noted!) quantities of test substance and reagents. Do not contaminate stock solutions shared by the group. Note results - colour and intensity - separately from conclusions.
- ☆ **Refinement :** Filter the contents of the tube in which you have been testing glucose (using paper previously labelled with pencil). Consider the colour of the filtrate, and accordingly modify the quantities of reagents used. Repeat filtration process. Explain your conclusions in your write-up.
- ☆ Dilute the sugar solutions (say x10, x100 etc), and carry out Benedict's test on suitable samples (i.e. amounts same as before). Explain the dilution technique you used in your write-up. What does this tell you about the sensitivity of the test?
- ☆ Using sucrose, carry out the test for non reducing sugars. Having obtained a negative result from ordinary benedict's test, a second test is performed after hydrolysing it to reducing sugars (acid hydrolysis, followed by neutralisation), then performing Benedict's test:- Add ... drops/ml HCl - leave for ... minutes at room temperature (... °C) or ... mins at 100 °C - add ... drops/ml NaOH (same strength) - check neutrality with litmus (colour ......) or universal indicator paper (colour .....)
- ☆ Test solution X using a combination of the above tests.
- ☆ Test starch solution using iodine solution (I/KI). You could use drops in a recessed tile, but for sensitivity purposes it is probably best to use boiling tubes and drops of iodine. - Dilute starch x10, x100, etc. and repeat

(standardised) test. - Warm the last tubes - *i.e.* the least blue coloured (100 °C waterbath)), & observe colour. - Cool (the lower end of) the tube under the tap. - Conclusion?

GLOSSARY

CATALYST: A substance that speeds up a chemical reaction without participating in it. Catalysts, of which enzymes are a good example, thus are not consumed in the reaction.

CELLULOSE: A polysaccharide, made from units of glucose, that is the principal material in the cell walls of plants. Cellulose also is found in natural fibers, such as cotton, and is used as a raw material in manufacturing such products as paper.

COMPLEX CARBOHYDRATE: A disaccharide, polysaccharide, or oligosaccharide. Also called a complex sugar.

DEXTROSE: Another name for glucose.

DISACCHARIDE: A double sugar, composed of two monosaccharides. Exam ples of disaccharides include the isomers sucrose, maltose, and lactose.

ENZYME: A protein material that speeds up chemical reactions in the bodies of plants and animals.

FRUCTOSE: Fruit sugar, a monosaccharide that is an isomer of glucose.

GALACTOSE: A monosaccharide and isomer of glucose. Less soluble and sweet than glucose, galactose usually appears in combination with other simple sugars rather than by itself.

GLUCOSE: A monosaccharide that occurs widely in nature and is the form in which animals usually receive carbohydrates. Also known as dextrose, grape sugar, and corn sugar.

GLYCOGEN: A white polysaccharide that is the most common form in which carbohydrates are stored in animal tissues, particularly muscle and liver tissues.

GUT: A term that refers to all or part of the alimentary canal, through which foods pass from the mouth to the intestines and wastes move from the intestines to the anus. Although the word is considered a bit crude in everyday life, physicians and bio logical scientists concerned with this part of the anatomy use it regularly.

ISOMERS: Two substances that have the same chemical formula but differ in chemical structure and therefore in chemical properties.

LACTOSE: Milk sugar. A disaccharide isomer of sucrose and maltose, lactose is the only major type of sugar that is produced from animal (i.e., mammal) rather than vegetable sources.

MALTOSE: A fermentable sugar generally formed from starch by the action of the enzyme amylase. Maltose is a disaccharide isomer of sucrose and lactose.

MONOSACCHARIDE: The simplest type of carbohydrate. Monosaccharides, which cannot be broken down chemically into simpler carbohydrates, also are known as simple sugars. Examples of monosaccharides include the isomers glucose, fructose, and galactose.

OLIGOSACCHARIDE: A carbohydrate containing a known, small number of monosaccharide units, typically between three and six. Compare with polysaccharide.

PHOTOSYNTHESIS: The biological conversion of light energy (that is, electromagnetic energy) from the Sun to chemical energy in plants. In this process, carbon dioxide and water are converted to carbohydrates and oxygen.

POLYSACCHARIDE: A carbohydrate composed of more than six monosaccharides. A polysaccharide sometimes is defined as containing two or more monosaccharides, but this definition does little to distinguish it from an oligosaccharide.

SACCHARIDE: A sugar.

SIMPLE SUGAR: monosaccharide, or simple carbohydrate.

STARCHES: Complex carbohydrates, without taste or odor, which are granular or powdery in physical form.

SUCROSE: Common table sugar (C12 H22 O11), a disaccharide formed from the bonding of a glucose molecule with a molecule of fructose. Sugar beets and cane sugar provide the principal natural sources of sucrose, which the average American is most likely to encounter in refined form as white, brown, or powdered sugar.

SUGARS: One of the three principal types of carbohydrate, along with starches and cellulose. Sugars can be defined as any of various water-soluble carbohydrates of varying sweetness. What we think of as "sugar" (i.e., table sugar) is actually sucrose.

CHAPTER- 5

AMINO ACID

5.1. PEPTIDE AND PROTEINS

Amino acids, peptides, and proteins are often referred to as related terms, yet they are different in their characteristics. Amino acids are the building blocks of both peptides and proteins. Amino acid is a small molecule that contains an amino group (-NH2) and a carboxylic acid group (-COOH), which are bound to a central carbon atom, with an additional hydrogen and a side chain (R- group). This side chain varies among all the amino acids; hence it determines the unique characters and the chemistry of each amino acid. Particular gene sequence is used to determine the amino acid sequence in both peptides and proteins.

5.1.1 Peptide

Peptides are made up with two or more amino acids, linked by peptide bonds and are present as linear chains. The length of a peptide is determined by the amount of amino acids in it. Usually the length of a peptide is less than approximately 100 amino acids.

Prefixes are used to describe the type of peptides in general terminology. For example, when a peptide is made of two amino acids, it is called dipeptide. Like that, three amino acids are combined to yield tripeptides, four amino acids are combined to yield tetrapeptides, etc. Apart from these types, there are oligopeptides (made up of 2-20 amino acids) and polypeptides, which have many peptides (less than 100). The most important characteristic features of peptides are determined by the amount and the sequence of amino acids. The primary function of the most peptides is to allow effective communication by carrying biochemical messages from one place to another in the body.

5.1.2 Protein

Proteins are the most diverse group of biological macromolecules. A protein is composed of one or more long unbranched chains called polypeptides and yet the building blocks of proteins are amino acids. The sequence of amino acid determines the major characteristics of a protein, while this amino acid sequence is defined by the particular gene sequence.

Usually proteins have stable three dimensional structures. These structures can be discussed in terms of a hierarchy of four levels; primary, secondary, tertiary and quaternary. The primary structure is the amino acid sequence of a protein. The secondary structure is produced by forming hydrogen bonds between two nearby amino acids, thus resulting structures called β-plated sheets, and coils called α-helices. The regions of secondary structure are then folded further in space to form the final three dimensional structures of the protein. Arrangement of the multiple polypeptides in space results the quaternary structure of a protein.

The primary structure of a protein is defined as the sequence of amino acids of which it is composed. This sequence ultimately determines the shape that the protein adopts, according to the spatial limitations on the arrangement of the atoms in the protein, the chemical properties of the component amino acid residues, and the protein's environment.

The peptide bonds that link amino acid residues in a polypeptide are formed in a condensation reaction between the acidic carboxyl group of one amino acid and the basic amino group of another amino acid. In the context of a peptide, the amide group (CO–NH) is referred to as the peptide group.

Crucial to an understanding of protein structure is a knowledge of the structure of the peptide bond. Linus Pauling, in the 1930s, used X-ray diffraction to examine the nature of the peptide bond formed between two amino acids. He reported that the peptide group (CO–NH) has a rigid planar structure. This structure is due to interactions between electrons of the double bond of the carbonyl group and those of the C–N bond such that the latter acquires partial (about 40%) double-bond properties.

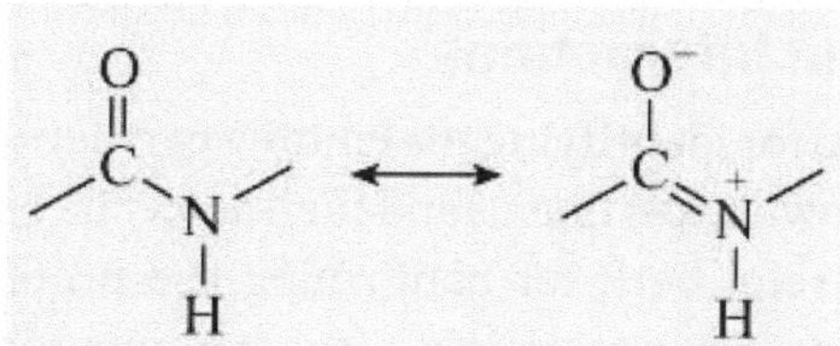

Fig Resonance interactions between electrons in C=O bond and C–N bond of peptide

This effect is an example of resonance which can be thought of as a sharing of electrons between bonds. Since single bonds between two atoms are longer than double bonds between the same two atoms, the lengths of the C–N and C=O bonds in the peptide group differ from those observed for these bonds in other contexts where resonance does not occur. Thus the partial double bond character of C–N in the peptide group means that this bond is shorter than would be predicted for a C–N single bond, whilst the C=O bond, having a partial single bond character due to resonance, is longer than would be predicted for a C=O double bond. The bond lengths in the peptide group are indicated in Fig. Compare the C–N bond of the peptide group with that between N and Cα (the C atom to which the amino group and carboxyl group are attached).

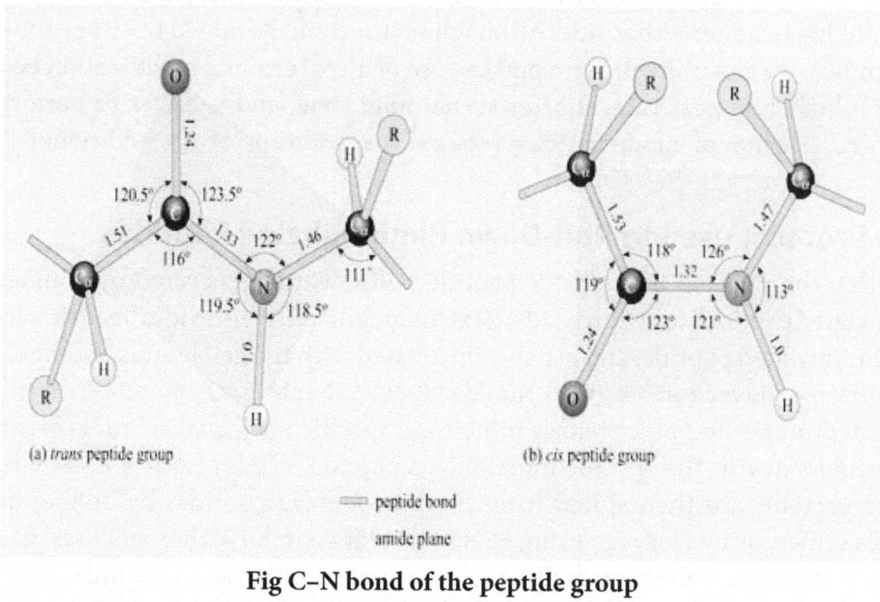

Fig C–N bond of the peptide group

Peptide bonds are in the trans conformation. However, cis forms can occur in peptide bonds that precede a proline residue. In such cases, the cis form is more stable than usual since the proline side-chain offers less of a hindrance. Nonetheless, cis peptide bonds occur only in approximately 10% of instances of peptide bonds preceding proline residues.

5.1.3 Peptide-Protein Interactions

The simplest method for identifying the binding partners of a peptide is to use it as bait in an affinity pull-down experiment, and then detect its binding proteins directly. Pull-down assays are useful both for confirming the protein-protein interactions that were predicted using other techniques (e.g., co-immunoprecipitation) and also as initial screening tools to identify novel protein-protein interactions. Synthetic peptides are commonly used to verify suspected protein-protein interactions by disrupting the binding competitively.

Biotinylated peptides, which contain a specific functional domain, and their corresponding control native peptides are immobilized to avidin-conjugated resins. Samples are incubated with the resins. Then wash the resins to remove any unbound proteins. The bound proteins are eluted and analyzed using SDS-PAGE. The amount of modified versus unmodified bound peptide can then be compared to identify specific functions.

Biotinylated peptides can be synthesized chemically with specific modifications at a purity >80%. Ideally, the peptides should be ~15–20 amino acids; the modification should be located in the center of the sequence, and there should be a minimum of 6–8 flanking residues either side. Although biotin could be added to either the N- or C-terminus, we generally recommend the use of an N-terminal modification because it has a higher success rate, shorter turnaround time, and is easier to perform. In contrast, C-terminal modifications require the addition of an additional lysine residue.

5.1.4 Protocol used to Pull-Down Biotin-labeled Peptides

First, the resin to be used for peptide pull-down is generated by conjugating biotinylated peptides to avidin beads. Next, immobilized streptavidin beads are loaded with biotinylated peptide, and are then incubated with the cell lysates obtained from cells that were lysed in 1% (v/v) Nonidet P-40, 150 mM NaCl, 50 mM Tris-HCl, pH 7.5 with protease and phosphatase inhibitors. Specifically, equal amounts of protein are incubated with the specific immobilized peptide of interest at 4°C for 6 h. The bound proteins are then eluted from the immobilized peptides by boiling in SDS sample buffer, or by cleavage using 50 mM dithiothreitol. Further analyses are then performed using eluates from pull-downs performed with the active and control bait peptides.

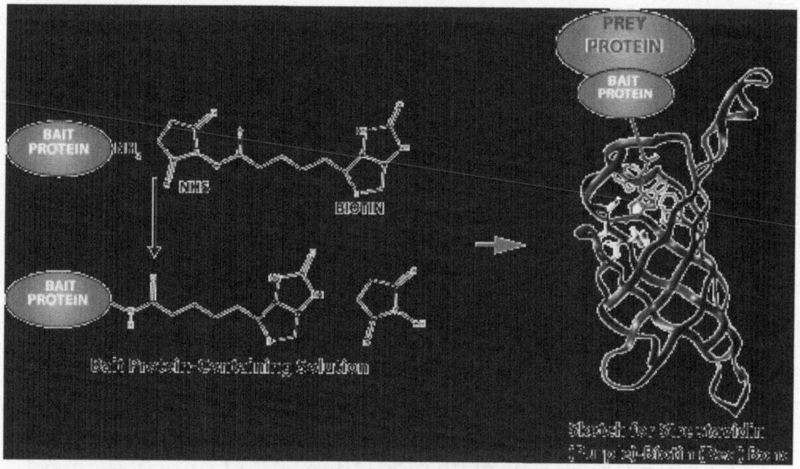

Fig streptavidin-biotin bond

The figure shows the force of streptavidin-biotin bond as they have an exceptionally high binding affinity, streptavidin-biotin and avidin-biotin pairs are two of the most prominent ligand-receptor pairs. Both proteins have a tetrameric structure; therefore, they can bind up to four ligands.

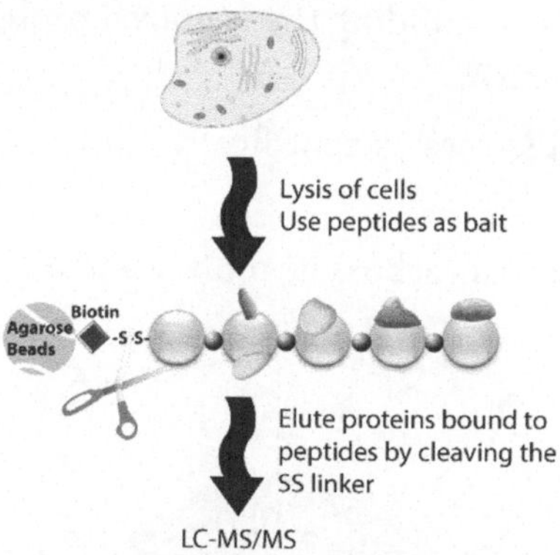

Peptides that correspond to specific regions that might be involved in protein-protein interactions can be synthesized. Proteins within the biological samples being analyzed then bind to the bait peptides, and are eluted, mixed, and analyzed. The eluted proteins are then digested using trypsin, and detected using mass spectrometry or SDS-PAGE. This technique is particularly useful for labeled cells. The reference described below provides an example of a method that used 13C-labeled cells. First, phosphorylated and non-phosphorylated peptides are synthesized separately. The 13C-labeled proteins were incubated with two related forms.

The data allow a larger peak intensity corresponding to the C13-labeled form to be identified from the phosphorylated bait peptide. Therefore, this novel technique uses synthetic unmodified and modified peptides to allow a proteomic screen to be performed to identify direct protein binding partners. Importantly, this screening method is unbiased, highly specific, and reproducible, and requires only the presence of the binding partner in the cell lysates. The resulting data are also selective because only a limited number of proteins are retrieved from the large number of proteins contained within whole cell lysates.

5.2 ESSENTIAL AND NON-ESSENTIAL AMINO ACIDS

5.2.1 Introduction

- Amino acids are the molecules which make up the proteins.
- Proteins are the building blocks of the body.
- In the body proteins comprise more than 50% of total body mass. While the rest are carbohydrates and fats.
- The proteins from food & other supplements we consume get broken down into amino-acids by the action of enzymes in the digestive system.
- These are then absorbed into the blood and supplied to whole body.
- They help the body in building the required protein in the cell by a process known as translation of m-RNA.
- These amino acids are scientifically differentiated as essential and nonessential amino acids.
- Every amino acid has a carboxylic moiety and an amino group.

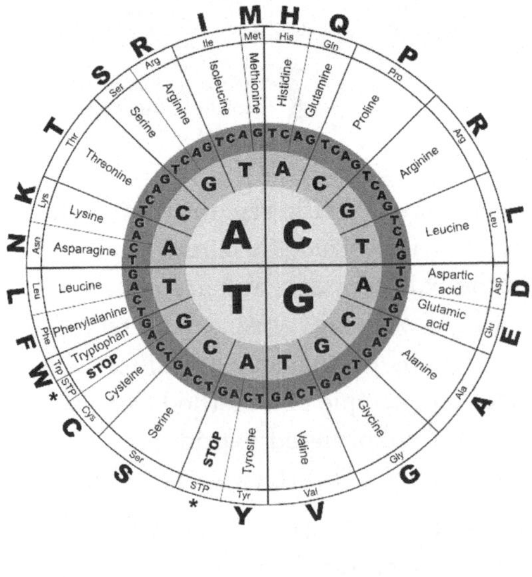

5.2.2 Essential Amino Acids

After carbohydrates, proteins are essential biomolecules for living beings. The main sources of protein are milk, cheese, pulses, fish etc. They are vital chemical substances which are essential for the growth and maintenance of the life. Proteins are high molecular mass complexes, biopolymers of amino acids. They are formed by the condensation polymerization of monomeric units; α-amino acids which bonded by peptide linkage.

There are total twenty α-amino acids which involve in the formation of proteins in living systems. All proteins are essentially made up of carbon, hydrogen, oxygen with nitrogen and sulfur. Some of the proteins contain non-metal like phosphorus, iodine as well as metals like iron, zinc, cobalt etc.

All proteins are made up of α-amino acids in which both functional groups; amino group and carboxyl group bonded on same carbon atom. Hence α-amino acids are building blocks of all proteins.

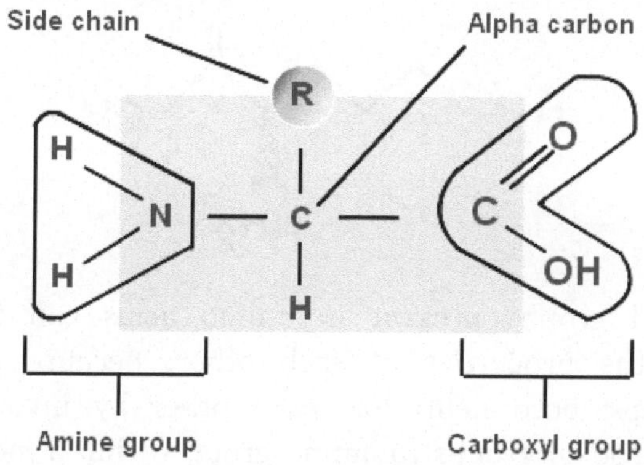

Fig α-amino acids

- ☆ In each αα-amino acid there are two functional group; amino group and carboxyl group bonded on same carbon atom.
- ☆ Excluding one amino acid, glycine, all other αα-amino acids are optically active as the α-carbon atom bonded with two different functional groups , one hydrogen atom and one alkyl group (-R).
- ☆ In optically inactive αα-amino acids; glycine, there are two hydrogen atoms on αα-carbon atom.
- ☆ In amino acids, both functional groups are of opposite nature.
- ☆ Amino group is basic in nature whereas carboxyl group is acidic in nature.
- ☆ Hence presence of both opposite functional groups makes amino acids natural in nature.

☆ If out of these functional groups, any one functional group is more than another one, molecules becomes acidic or basic.

☆ More than one amino group make amino acid basic in nature and called as basic amino acid. Like, Arginine contains total four amino groups with one carboxyl group.

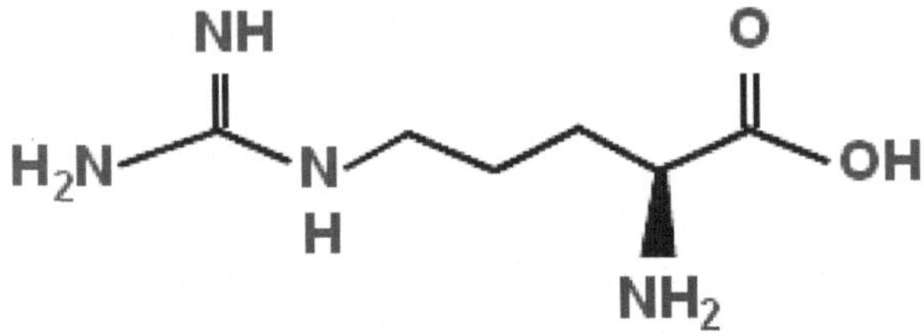

If more than one carboxyl groups in amino acid make molecule acidic like aspartic acid (HOOCCH(NH2)CH2COOH) contains two carboxyl groups with one amino acid.

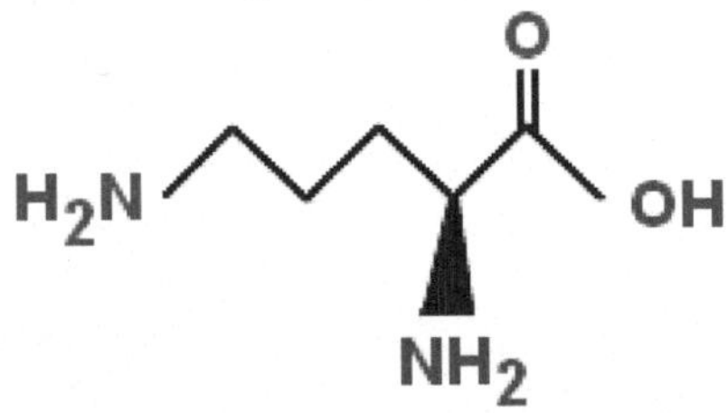

Both functional groups present in amino acids can interact with each other and affect the properties of each other. Because of opposite nature of functional groups; both neutralize each other by involving transfer of a proton from the carboxyl groups to amino group within a molecule. As a results amino acid exists as dipolar ion also called as zwitterion or internal salt.

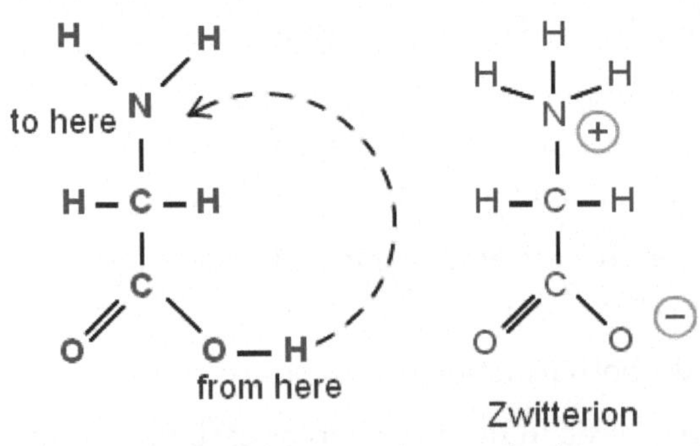

Fig zwitterion or internal salt

While in basic medium, the carboxyl group releases proton and forms carboxylate ion which move towards anode. At a certain pH both forms are in equilibrium, hence show no overall movement towards electrodes in electrolysis. That pH value is known as isoelectric point. It is a characteristic value for each amino acid.

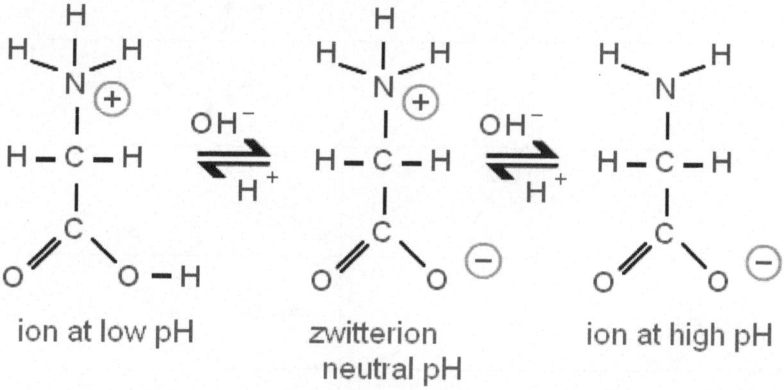

Out of the twenty amino acids, about ten amino acids cannot be synthesized by human body and must be supplied by food to body and called as essential amino acid or indispensable amino acids. These amino acids are required for the growth of the body and their deficiency causes diseases like kwashiorkor in which the water balance of living body get disturbed and some of the organs of the body become watery and bloated. Therefore essential amino acids must be supplied by human diet.

Other bio molecules like fat and carbohydrate can store in human body, while amino acid cannot store, hence continues supply of amino acids are necessary for growth and development and they must be present in the food of every day. Some of the essential amino acids are Isoleucine, Leucine, Trypophan, and Valine. Some essential amino acids synthesized in body but with a very low rate which is not sufficient for the necessary amount of needed amino acids.

For example, Arginine can be synthesized by mammalian cells but with very slow speed, hence insufficient for all requirements. Some amino acids are complementary to each other, hence one required for the growth of another one. Like Methionine is required to produce cysteine and phenyalanine is needed to form tyrosine.

All amino acids can be written as abbreviation of three words or one word abbreviation. Ten essential amino acids with their structure and abbreviations are as follows;

Essential amino acid	3-letter abbreviation	1-letter abbreviation	Structure	Polarity
Histidine	His	H	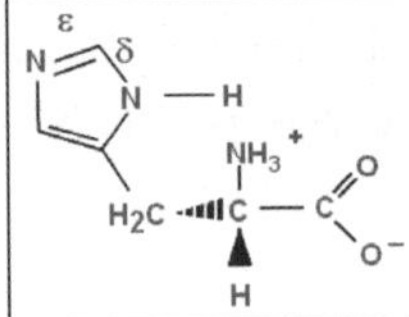	
Isoleucine	Ile	I	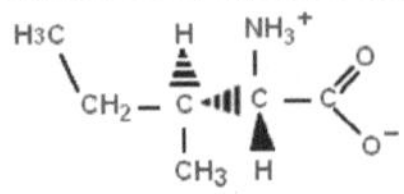	Non-polar
Leucine	Leu	L	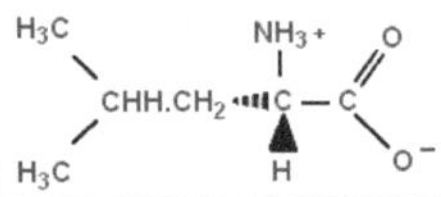	Non-polar
Lysine	Lys	K	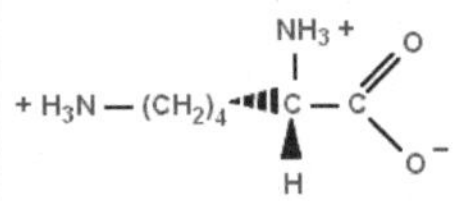	Polar
Methionine	Met	M	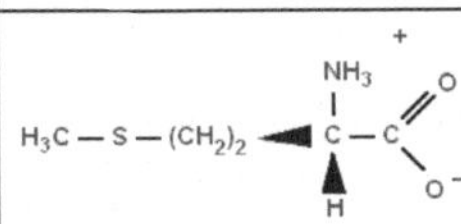	Non-polar
Threonine	Thr	T	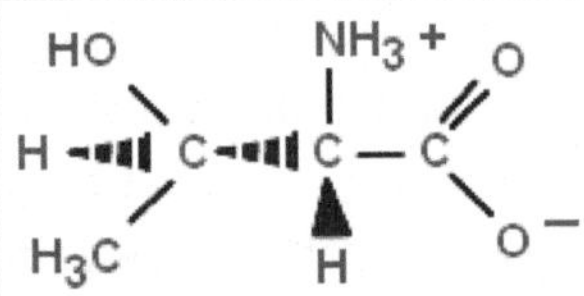	Polar
Tryptophan	Trp	W	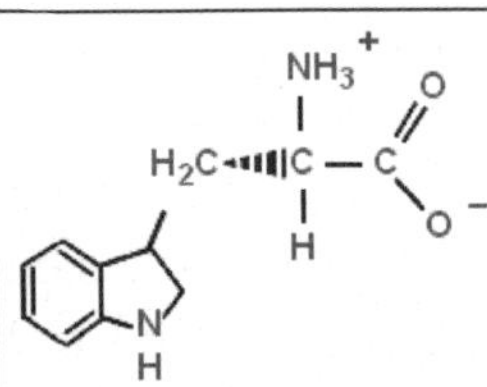	Non-polar
Valine	Val	V	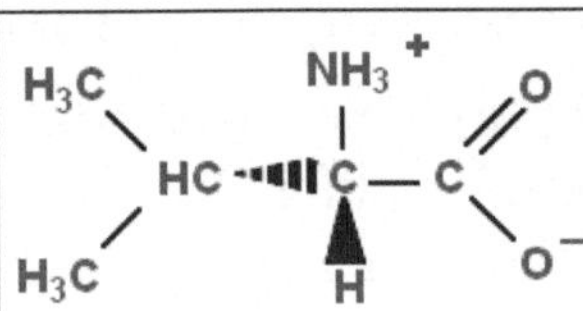	Non-polar

5.2.3 Non Essential Amino Acids

Some of the amino acids can readily be synthesised in the body, from more or less common metabolic intermediates. These are known as non-essential, or dispensible, amino acids, since there is no need for them to be provided in the diet. As long as there is enough total protein in the diet, these amino acids can be synthesised.

The other type is the non-essential amino acid, 11 of which exist and are synthesized by the body. Thus, although they are an important part of building proteins, they do not need to be included in an everyday diet. Eight of these non-essential acids are also known as "conditional," meaning that the body may not be capable of producing enough of them when presented with substantial stress or illness.

Nonessential amino acid	Structure	Properties
Alanine	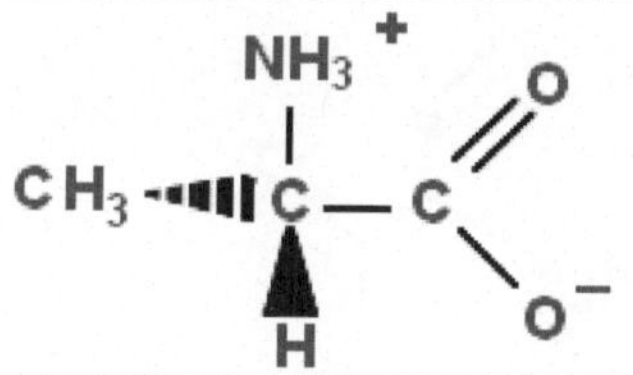	Non-polar, optically active amino acid.
Arginine*	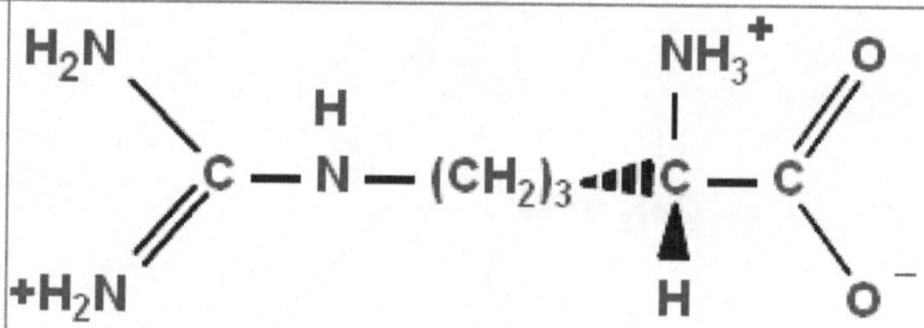	Polar , basic amino acid with two amino group. Become essential amino acid in some conditions.
Asparagine	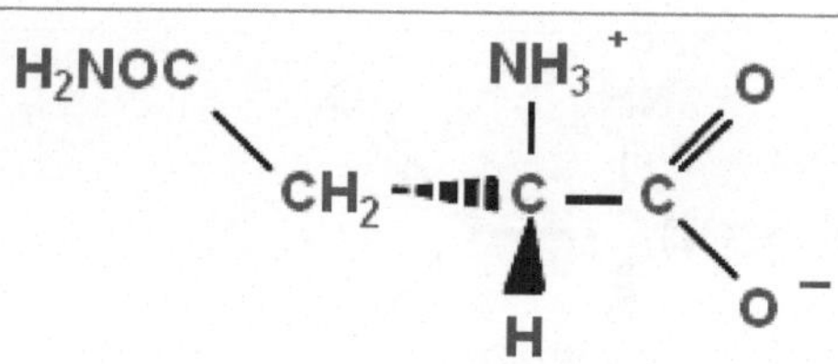	Amino acid with amide group.
Aspartic acid	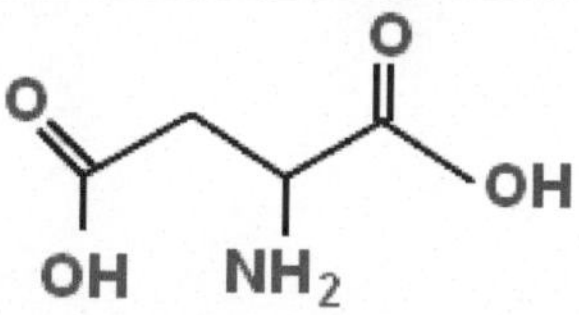	Acidic polar amino acid with two carboxyl group

There are ten amino acids which can be synthesized in human body which called as non-essential or dispensable amino acids. Since they can be synthesized in human body, hence they regular supply through diet is not essential. Nonessential is a slightly misleading label because these amino acids actually fill essential roles, but since they're synthesized by your body, they're not an essential part of your diet. Of the 11 nonessential amino acids, eight are called conditional amino acids. When you're sick or under significant stress, your body may not be able to produce enough of these amino acids to meet your needs. The list of conditional amino acids includes arginine, glutamine, tyrosine, cysteine, glycine, proline, serine and ornithine.

Cysteine*	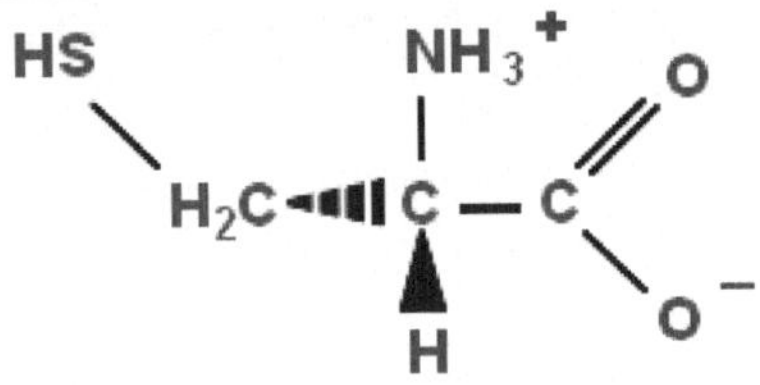	Amino acid with sulphur ,non-polar.
Glutamic acid	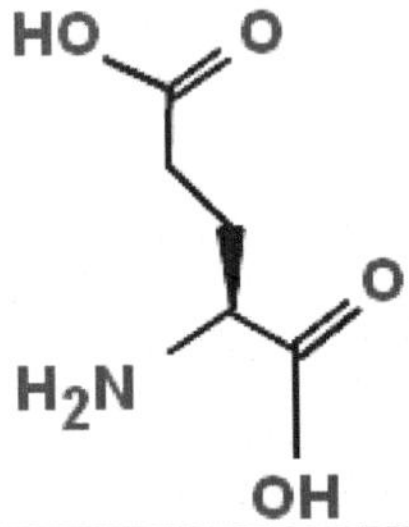	Polar , acidic amino acid wit two carboxyl groups.
Glutamine*	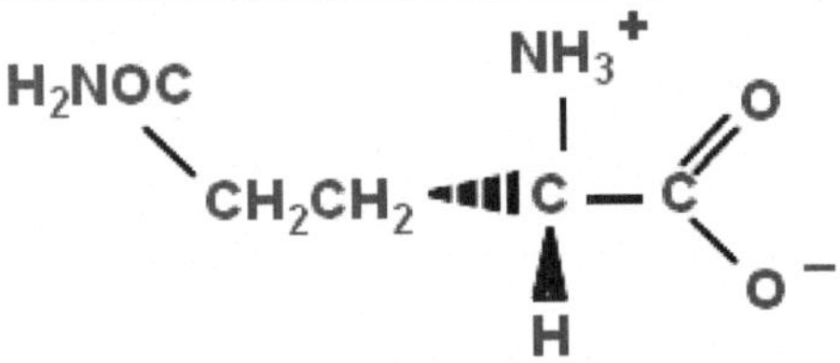	Become essential amino acid in some conditions.
Glycine	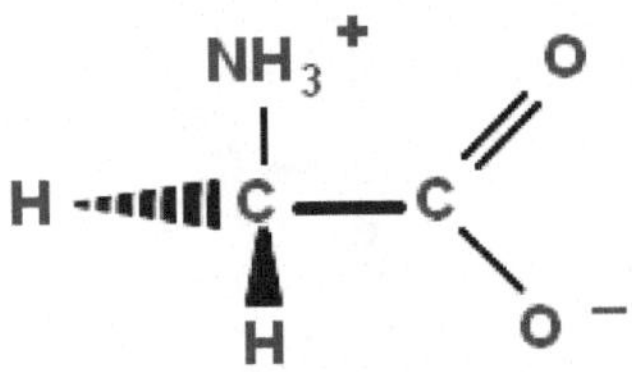	Optically inactive amino acid.

Proline*	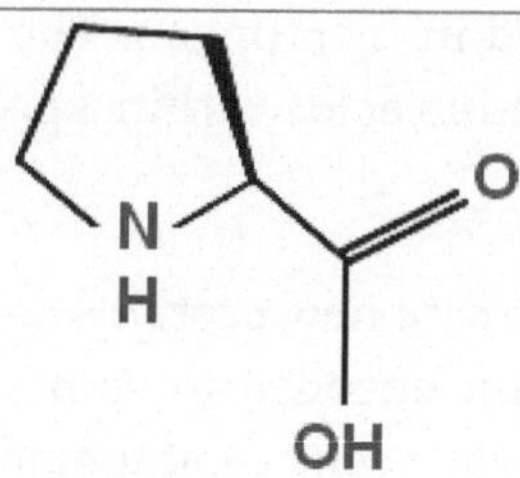	Only amino acid with imine group. Become essential amino acid in some conditions.
Selenocysteine*	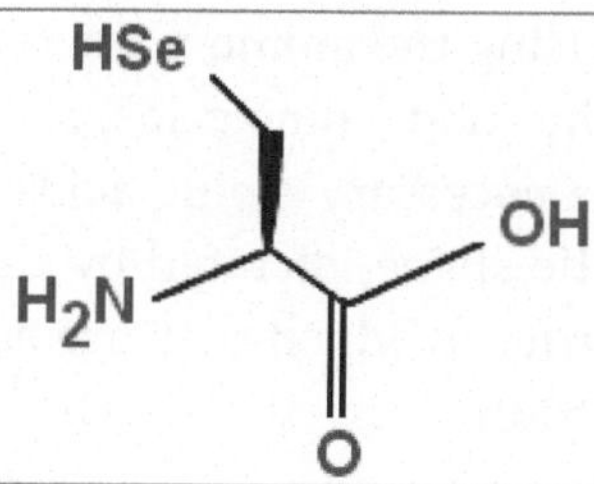	Amino acid with Seleium , present in many enzymes.
Serine*	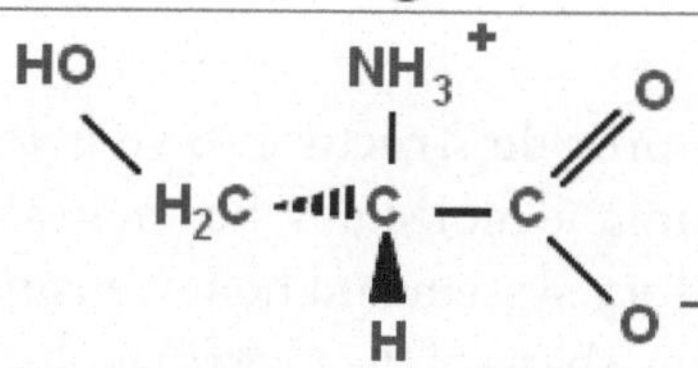	Amino acid with hydroxy group. Become essential in some conditions.
Taurine*	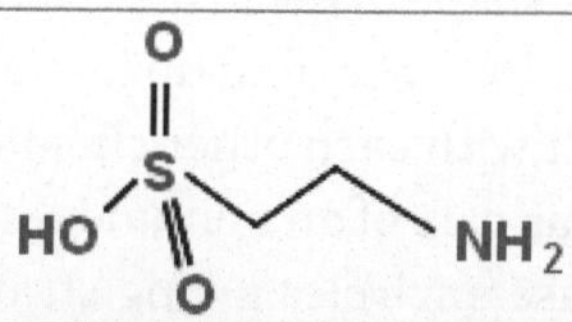	An amino acid with sulfonic group, present in some animal tissues.

5.3 AMINO ACIDS BUILDING BLOCKS OF PROTEINS

5.3.1 Building Blocks: Amino Acids

The basic building block of a protein is called an amino acid. There are 20 amino acids in the proteins you eat and in the proteins within your body, and they link together to form large protein molecules. The variety of ways they mix and match allows for the great diversity of proteins in nature. Understanding the structure of amino acids and how they relate to the proteins they create can help you appreciate how critical these nutrients are in your diet.

Although each of the 20 amino acids is different, the basic structure follows a similar pattern. Amino acids all contain a carbon atom in the middle of the molecule, which is surrounded by three chemical groups. Two of these chemical groups, a carboxyl group and an amino group, are also the same in all amino acids. The third group, however, differentiates each amino acid from the others. These R-groups can be as simple as a single hydrogen atom, as in the amino acid glycine, or as complex as the double-ring structure found in tryptophan. The chemical differences in the R-groups affect the interplay of amino acids within a protein.

5.3.2 Role in Protein Synthesis

Your cells have the ability to create new proteins as your body needs them. They follow a kind of chemical blueprint encoded in your genes that tells the protein-manufacturing machinery within your cells what amino acids are in the protein and in what order they need to be. Using the amino acids that result from digestion of the proteins you eat, your cells then add amino acids, one by one, to a growing chain of these molecules. When all the necessary amino acids have been added, the new protein chain folds into a specific shape, dictated by the chemical nature of each individual amino acid and the order in which they are assembled. The shape of the protein impacts its ultimate function.

5.3.3 Functions

The proteins in your body provide structure to your muscles, organs, skin and hair. They work as enzymes, immune molecules, hormones and cellular messengers. They also support your circulatory system and help transmit nerve impulses. These different functions depend on the shape of the protein molecule, which is determined by its amino acids. Antibodies, for instance, fold up into a Y shape because of the way their specific amino acids interact with each other chemically. The top portion of the Y can then trap foreign invaders as part of an immune response. As another example, doughnut-shaped DNA polymerase encircles a long strand of DNA as it assists with protein synthesis.

5.3.4 Getting Amino Acids From Your Diet

Of the 20 amino acids in the foods you eat, your body can manufacture 11 of them if necessary. The other nine – the essential amino acids – you can get only if you include them in your diet. Without them, you can't create all the proteins your body needs for good health. Every animal source of protein gives you all the essential amino acids, and certain combinations of plant proteins can provide you with these nutrients as well. The essential amino acids missing in whole grains, for example, are available in legumes, and consuming both together – such as split pea soup with a whole-wheat roll – supplies you with a complete protein containing all essential amino acids.

5.4 CLASSIFICATION OF AMINO ACIDS

Most of the amino acids we are familiar with are known as standard amino acids. These are amino acids that are formed from universal genetic coding. Standard amino acids (also known as canonical amino acids) are protein 'building-blocks'. These are the amino acids that together with DNA help organism to form and function. Non-standard amino acids do not help to build proteins. There are only three exceptions to this rule. Selenocysteine, pyrrolysine, and N-formylmethionine are non-standard amino acids that can still be naturally incorporated into proteins. They do this by exploiting genetic code that has not been used or in a certain position within an amino acid chain. Non-standard amino acids can also be modified forms of standard amino acids. Many non-standard amino acids are used in genetic engineering.

5.4.1 Depending upon the Charge

Amino acids can be broadly classified into three major groups:

- ☆ Neutral
- ☆ Acidic
- ☆ Basic.

1. Neutral amino acids

- ☆ Those amino acids that do not contain any charge on the 'R' group.
- ☆ They are further classified into the following categories:
- ☆ **Aliphatic:** Those amino acids whose 'R' group contains a chain of carbon atoms—Gly, Ala, Ser, Thr, Val, Leu, lie, Asn, Gin.
- ☆ **Aromatic:** Those amino acids whose 'R' group has a benzene ring—Phe, Tyr, Trp.
- ☆ **Heterocyclic:** The "R" group has a heterocylic ring, i.e., any of the ring structures which contain different atoms—Pro, His.
- ☆ **Sulphur containing:** Those amino acids which contain a sulphur atom-Cys, Met.

2. Acidic amino acids

- ☆ Those amino acids that contain a negative charge or an acidic group-Asp, Glu.

3. Basic amino acids

- ☆ Those amino acids that contain a positive charge or a basic group-Arg, Lys and His.

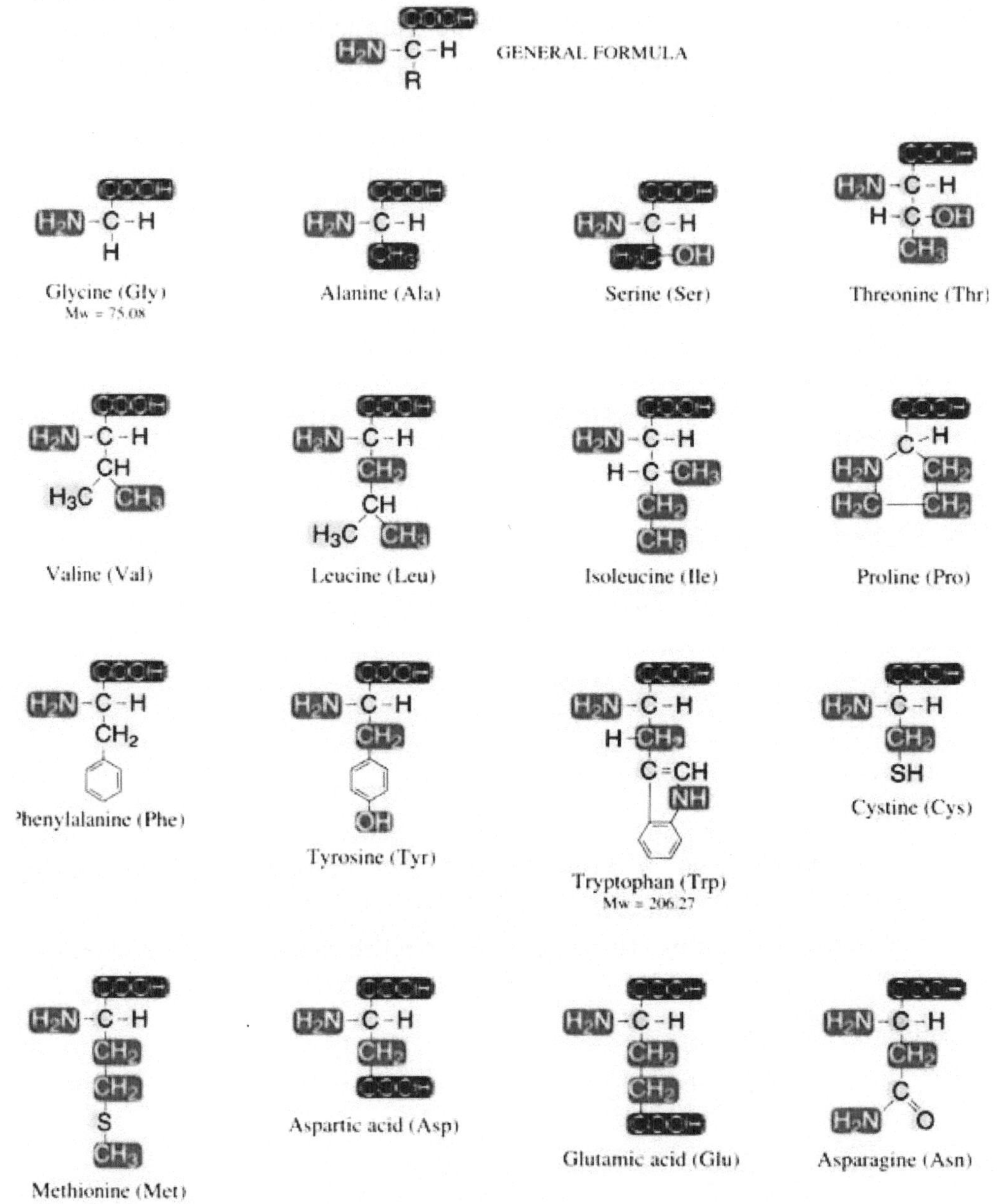

5.4.2 Depending upon the Solubility in Water

☆ The amino acids can also be grouped into two different categories, depending upon their solubility in water. They are—

☆ Hydrophobic amino acids: Amino acids insoluble in water are known as hydrophobic amino acids. They are—Ala, Val, Leu, lie, Pro, Met, Phe, Trp.

☆ Hydrophilic amino acids:

☆ Amino acids soluble in water are known as hydrophilic amino acids. They are—Gly, Ser, Thr, Cys, Tyr, Asp, Asn, Glu, Gin, Lys, Arg, His.

5.4.3 Depending upon their Nutritional Requirements:

The amino acids are classified into two groups.

Essential amino acids

Are those which cannot be synthesized by the human body and hence they should be taken through the diet. There are 10 essential amino acids. Among these amino acids, arginine and histidine are known as semi-essential amino acids.

M – Methionine	**V** – Valine	**P** – Phenylalanine
A – Arginine	**I** – Isoleucine	**H** – Histidine
T – Threonine	**L** – Leucine	**Ly** – Lysine
T – Tryptophan		

Non-essential amino acids

These acids are those that can be synthesized in the human body and are not required in the diet. These include gly, ala, ser, pro, tyr, cys, asp, asn, glu, gin.

5.5 REACTIONS OF AMINO ACIDS:

5.5.1. Zwitter ions: Amino acids have an acidic group (—COOH group), i.e., a proton, donor. They also have a basic group (—NH2 group), i.e., a proton, acceptor. A compound capable of both donating and accepting protons and thus able to act either as an acid or a base is known as amphoteric molecule. Amino acids have both anions and cations in solution and such compounds are called zwitter ions.

5.5.2. Isoelectric pH (pH1):

The pH at which the positive charge on the amino acid (or any other molecule) is equal to the negative charges, is known as isoelectric pH. At this pH the net charge will be zero and hence it does not move either to positive (anode) or to negative (cathode) electrode, when subjected to an electric field. At pH1 all the molecules exists in zwitter ion form.

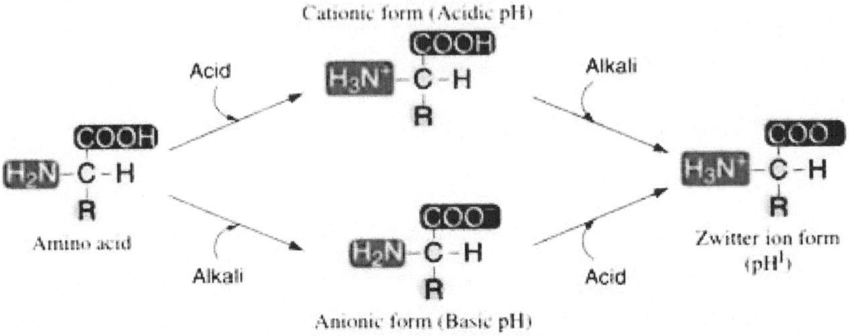

5.5.3. Reactions due to amino group:

(a) Ninhydrin test: This test identifies or detects amino acids. If amino acids are heated with ninhydrin, they form a purple blue coloured compound, which is measured colorimetrically.

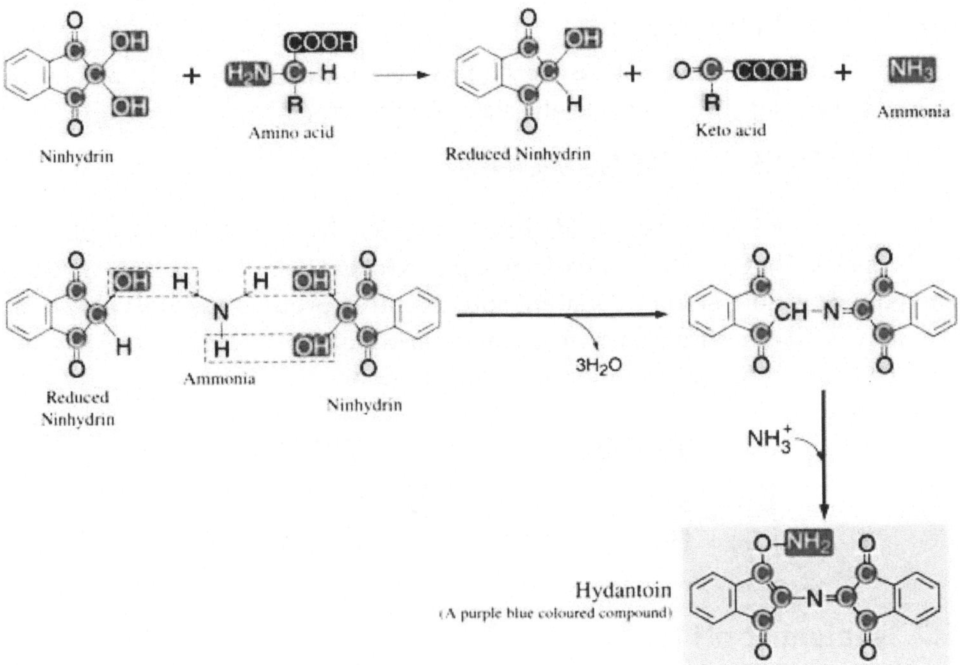

5.6. STRUCTURE AND PROPERTIES OF AMINO ACIDS

5.6.1 Structure

Proteins are polymers of the bifunctional monomer, amino acids. The twenty common naturally-occurring amino acids each contain an a-carbon, an a-amino group, an a-carboxylic acid group, and an a-side chain or side group. These side chains (or R groups) may be either nonpolar, polar and uncharged, or charged, depending on the pH and pKa of the ionizable group. Two other amino acids occasionally appear in proteins. One is selenocysteine, which is found in Arachea, eubacteria, and animals.

Another just recently found is pyrrolysine, found in Arachea. Shultz et al. have gone one step farther. They have engineered bacterial to incorporate two new amino acids, O-methyl-tyrosine and p-aminophenylalanine. More recently, they (Chin et al.) have engineered the yeast strain Saccharomyces cerevisiae to incorporate five new unnatural amino acid (using the TAG nonsense codon and new, modified tRNA and tRNA synthetases) with keto groups that allow chemical modifications to the protein.

Amino Acid	3 Letter Code	1 Letter Code	Structure	pKa1(α-carboxyl)	pka2 (α-amino)	pka3 (side chain)
Nonpolar Side Chains						
Glycine	Gly	G	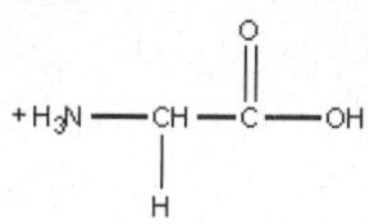	2.35	9.78	.
Alanine	Ala	A	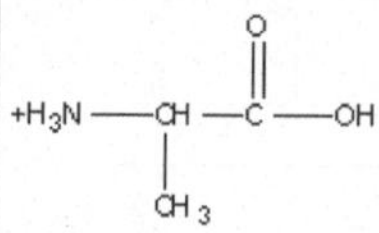	2.35	9.87	.
Valine	Val	V	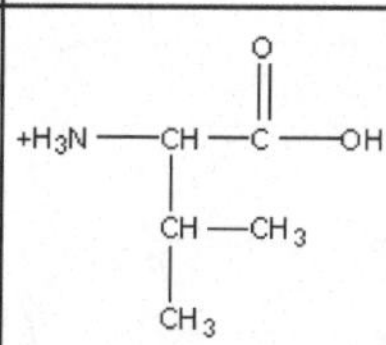	2.29	9.74	.
Leucine	Leu	L	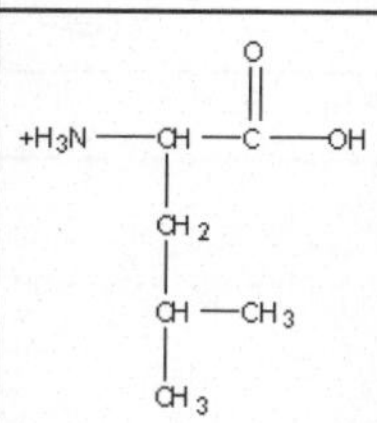	2.33	9.74	.

Isoleucine	Ile	I	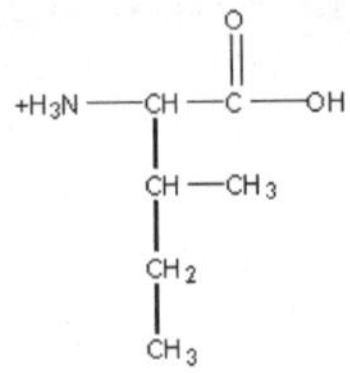	2.32	9.76	.
Methionine	Met	M	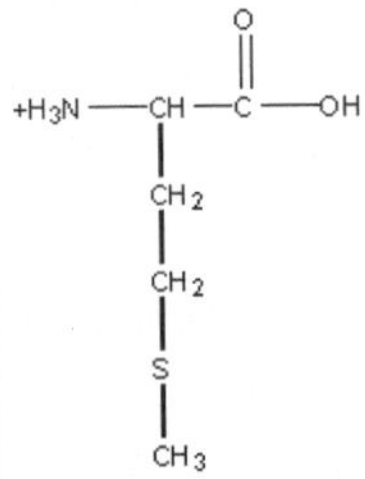	2.13	9.28	.
Proline	Pro	P	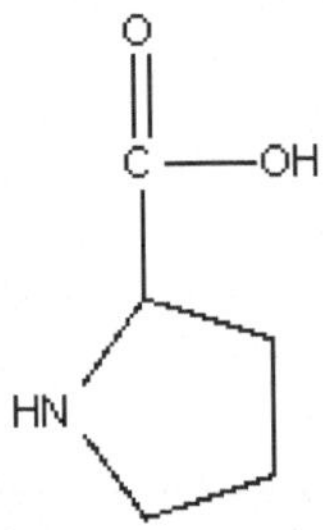	1.95	10.64	.
Phenylalanine	Phe	F	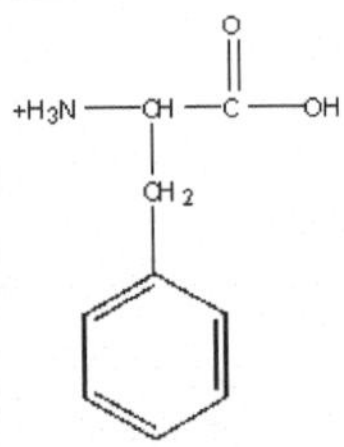	2.20	9.31	.
Tryptophan	Trp	W	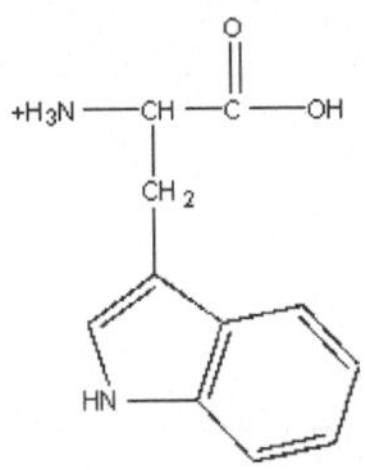	2.46	9.41	.
Uncharged Polar Side Chains						
Serine	Ser	S	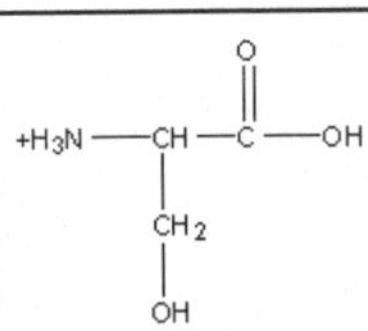	2.19	9.21	.

Threonine	Thr	T	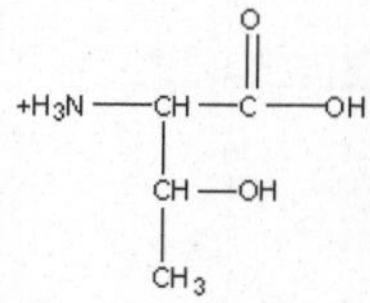	2.09	9.10	.
Asparagine	Asn	N	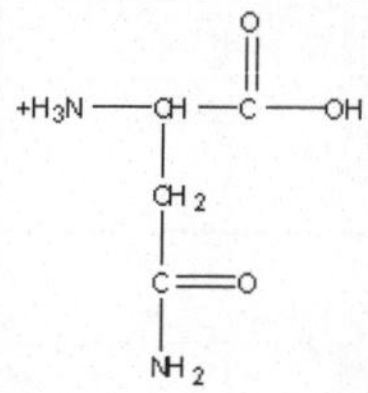	2.14	8.72	.
Glutamine	Gln	Q	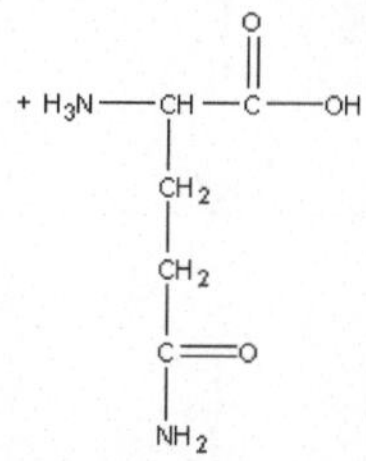	2.17	9.13	.
Tyrosine	Tyr	Y	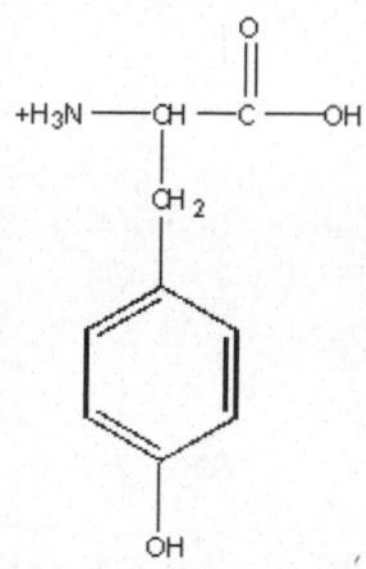	2.20	9.21	10.46
Cysteine	Cys	C	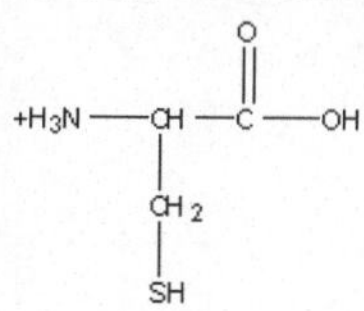	1.92	10.70	8.37
Charged Polar Side Chains10.54						
Lysine	Lys	K	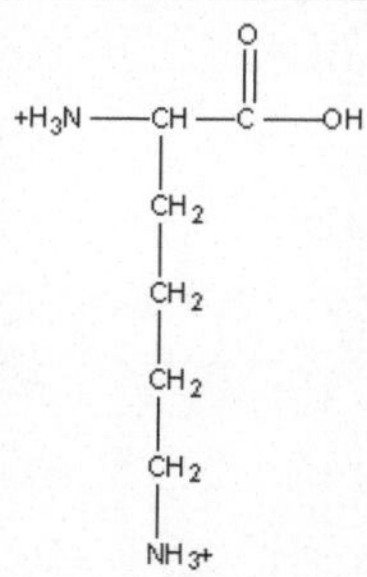	2.16	9.06	10.54

Arginine	Arg	R	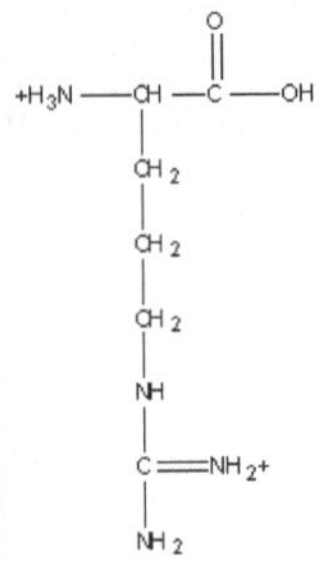	1.82	8.99	12.48
Histidine	His	H	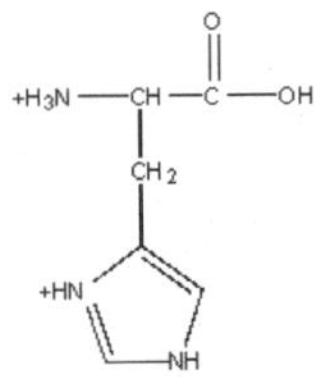	1.80	9.33	6.04
Aspartic Acid	Asp	D	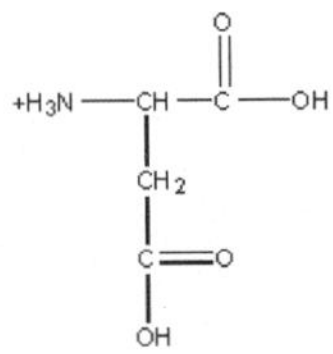	1.99	9.90	3.90
Glutamic Acid	Glu	E	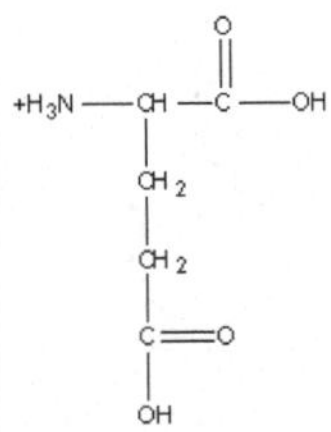	2.10	9.47	4.07

5.6.2 Properties

Name	3-Letter Symbol	1-Letter Symbol	Molecular Weight	Molecular Formula	Residue Formula	Residue Weight (-H_2O)	pKa[1]	pKb[2]	pKx[3]	pl[4]
Alanine	Ala	A	89.10	$C_3H_7NO_2$	C_3H_5NO	71.08	2.34	9.69	–	6.00
Arginine	Arg	R	174.20	$C_6H_{14}N_4O_2$	$C_6H_{12}N_4O$	156.19	2.17	9.04	12.48	10.76
Asparagine	Asn	N	132.12	$C_4H_8N_2O_3$	$C_4H_6N_2O_2$	114.11	2.02	8.80	–	5.41
Aspartic acid	Asp	D	133.11	$C_4H_7NO_4$	$C_4H_5NO_3$	115.09	1.88	9.60	3.65	2.77
Cysteine	Cys	C	121.16	$C_3H_7NO_2S$	C_3H_5NOS	103.15	1.96	10.28	8.18	5.07
Glutamic acid	Glu	E	147.13	$C_5H_9NO_4$	$C_5H_7NO_3$	129.12	2.19	9.67	4.25	3.22
Glutamine	Gln	Q	146.15	$C_5H_{10}N_2O_3$	$C_5H_8N_2O_2$	128.13	2.17	9.13	–	5.65
Glycine	Gly	G	75.07	$C_2H_5NO_2$	C_2H_3NO	57.05	2.34	9.60	–	5.97
Histidine	His	H	155.16	$C_6H_9N_3O_2$	$C_6H_7N_3O$	137.14	1.82	9.17	6.00	7.59
Hydroxyproline	Hyp	O	131.13	$C_5H_9NO_3$	$C_5H_7NO_2$	113.11	1.82	9.65	–	–
Isoleucine	Ile	I	131.18	$C_6H_{13}NO_2$	$C_6H_{11}NO$	113.16	2.36	9.60	–	6.02
Leucine	Leu	L	131.18	$C_6H_{13}NO_2$	$C_6H_{11}NO$	113.16	2.36	9.60	–	5.98
Lysine	Lys	K	146.19	$C_6H_{14}N_2O_2$	$C_6H_{12}N_2O$	128.18	2.18	8.95	10.53	9.74
Methionine	Met	M	149.21	$C_5H_{11}NO_2S$	C_5H_9NOS	131.20	2.28	9.21	–	5.74
Phenylalanine	Phe	F	165.19	$C_9H_{11}NO_2$	C_9H_9NO	147.18	1.83	9.13	–	5.48
Proline	Pro	P	115.13	$C_5H_9NO_2$	C_5H_7NO	97.12	1.99	10.60	–	6.30
Pyroglutamatic	Glp	U	139.11	$C_5H_7NO_3$	$C_5H_5NO_2$	121.09	–	–	–	5.68
Serine	Ser	S	105.09	$C_3H_7NO_3$	$C_3H_5NO_2$	87.08	2.21	9.15	–	5.68
Threonine	Thr	T	119.12	$C_4H_9NO_3$	$C_4H_7NO_2$	101.11	2.09	9.10	–	5.60
Tryptophan	Trp	W	204.23	$C_{11}H_{12}N_2O_2$	$C_{11}H_{10}N_2O$	186.22	2.83	9.39	–	5.89
Tyrosine	Tyr	Y	181.19	$C_9H_{11}NO_3$	$C_9H_9NO_2$	163.18	2.20	9.11	10.07	5.66
Valine	Val	V	117.15	$C_5H_{11}NO_2$	C_5H_9NO	99.13	2.32	9.62	–	5.96

[1] pKa is the negative of the logarithm of the dissociation constant for the -COOH group.
[2] pKb is the negative of the logarithm of the dissociation constant for the -NH_3 group.
[3] pKx is the negative of the logarithm of the dissociation constant for any other group in the molecule.
[4] pl is the pH at the isoelectric point.

5.7 PEPTIDE BONDS

If the amine and carboxylic acid functional groups in amino acids join together to form amide bonds, a chain of amino acid units, called a peptide, is formed. A simple tetrapeptide structure is shown in the following diagram. By convention, the amino acid component retaining a free amine group is drawn at the left end (the N-terminus) of the peptide chain, and the amino acid retaining a free carboxylic acid is drawn on the right (the C-terminus). As expected, the free amine and carboxylic acid functions on a peptide chain form a zwitterionic structure at their isoelectric pH.

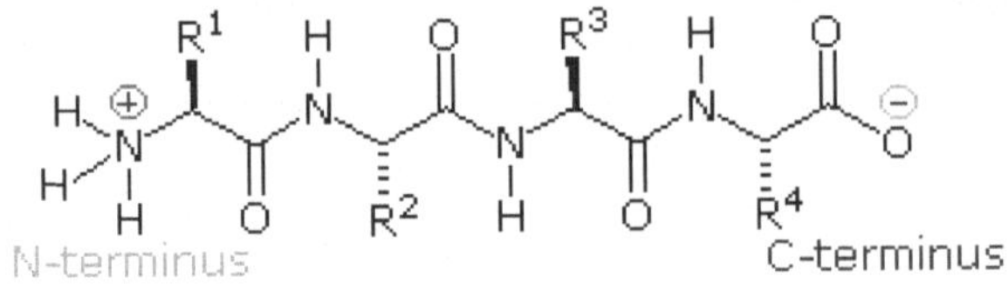

Fig amide bonds

The conformational flexibility of peptide chains is limited chiefly to rotations about the bonds leading to the alpha-carbon atoms. This restriction is due to the rigid nature of the amide (peptide) bond. As shown in the following diagram, nitrogen electron pair delocalization into the carbonyl group results in significant double bond character between the carbonyl carbon and the nitrogen. This keeps the peptide links relatively planar and resistant to conformational change. The color shaded rectangles in the lower structure define these regions, and identify the relatively facile rotations that may take place where the corners meet (i.e. at the alpha-carbon). This aspect of peptide structure is an important factor influencing the conformations adopted by proteins and large peptides.

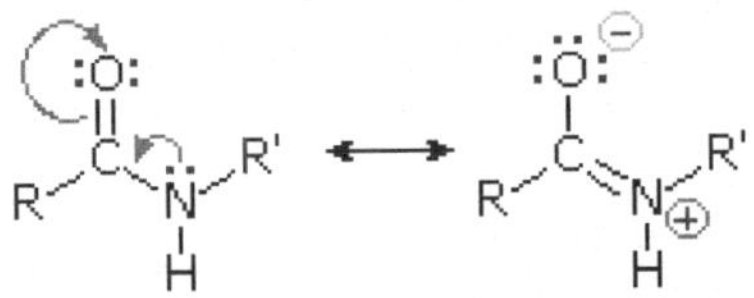

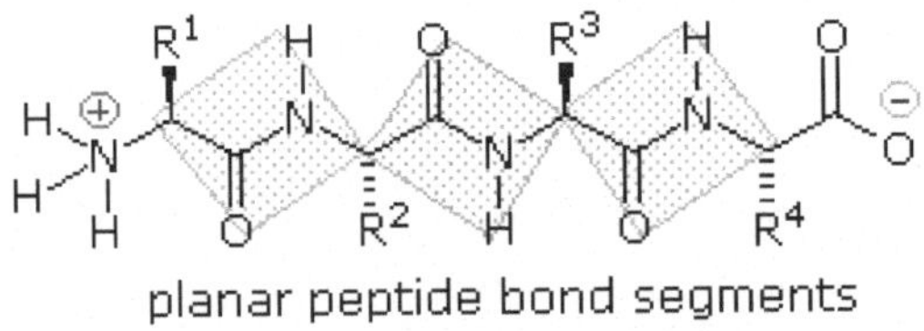

Fig peptide structure

5.7.1 Primary Structure of Peptides

Because the N-terminus of a peptide chain is distinct from the C-terminus, a small peptide composed of different aminoacids may have a several constitutional isomers. For example, a dipeptide made from two different amino acids may have two different structures. Thus, aspartic acid (Asp) and phenylalanine (Phe) may be combined to make Asp-Phe or Phe-Asp, remember that the amino acid on the left is the N-terminus. The methyl ester of the first dipeptide (structure on the right) is the artificial sweetener aspartame, which is nearly 200 times sweeter than sucrose. Neither of the component amino acids is sweet (Phe is actually bitter), and derivatives of the other dipeptide (Phe-Asp) are not sweet.

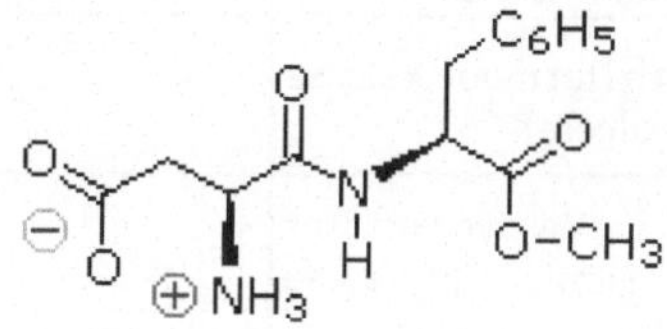

Asp-Phe-OCH3
Aspartame

A tripeptide composed of three different amino acids can be made in 6 different constitutions, and the tetrapeptide shown above (composed of four different amino acids) would have 24 constitutional isomers. When all twenty of the natural amino acids are possible components of a peptide, the possible combinations are enormous. Simple statistical probability indicates that the decapeptides made up from all possible combinations of these amino acids would total 2010.

Natural peptides of varying complexity are abundant. The simple and widely distributed tripeptide glutathione (first entry in the following table), is interesting because the side-chain carboxyl function of the N-terminal glutamic acid is used for the peptide bond. An N-terminal glutamic acid may also close to a lactam ring, as in the case of TRH (second entry). The abbreviation for this transformed unit is pGlu (or pE), where p stands for "pyro" (such ring closures often occur on heating). The larger peptides in the table also demonstrate the importance of amino acid abbreviations, since a full structural formula for a nonapeptide (or larger) would prove to be complex and unwieldy. The ten peptides listed in this table make use of all twenty common amino acids. Note that the C-terminal unit has the form of an amide in some cases (e.g. TRH, angiotensin & oxytocin).

Natural Peptides		
Name (residues)	**Source or Function**	**Amino Acid Sequence**
Glutathione (3)	Most Living Cells (stimulates tissue growth)	$(+)H_3NCH(CO_2(-))CH_2CH_2CONHCH(CH2SH)CONHCH_2CO_2H$ γ-Glu-Cys-Gly (or γECG)
TRH (3)	Hypothalmic Neurohormone (governs release of thyrotropin)	
Angiotensin II (8)	Pressor Agent (acts on the adrenal gland)	Asp-Arg-Val-Tyr-Ile-His-Pro-PheNH2 (or DRVYIHPFNH2)
Bradykinen (9)	Hypotensive Vasodilator (acts on smooth muscle)	Arg-Pro-Pro-Gly-Phe-Ser-Pro-Phe-Arg (or RPPGFSPFR)
Oxytocin (9)	Uterus-Contracting Hormone (also stimulates lactation)	

Natural Peptides		
Name (residues)	**Source or Function**	**Amino Acid Sequence**
Somatostatin (14)	Inhibits Growth Hormone Release (used to treat ulcers)	
Endothelin (21)	Potent Vasoconstrictor (structurally similar to some snake venoms)	<=”” td=””>
Melittin (26)	Honey Bee Venom (used to treat rheumatism)	Gly-Ile-Gly-Ala-Val-Leu-Lys-Val-Leu-Thr-Thr-Gly-Leu-Pro~ ~Ala-Leu-Ile-Ser-Trp-Ile-Lys-Arg-Lys-Arg-Gln-GlnNH2 (or GIGAVLKVLTTGLPALISWIKRKRQQNH2)
Glucagon (29)	Hyperglycemic Factor (used as an anti-diabetic)	His-Ser-Gln-Gly-Thr-Phe-Thr-Ser-Asp-Tyr-Ser-Lys-Tyr-Leu-Asp~ ~Ser-Arg-Arg-Ala-Gln-Asp-Phe-Val-Gln-Trp-Leu-Met-Asn-Thr (or HSQGTFTSDYSKYLDSRRAQDFVQWLMNT)
Insulin (51)	Pancreatic Hormone (used in treatment of diabetes)	<=”” td=””>

Partial hydrolysis will produce a mixture of shorter peptides and some amino acids. If the primary structures of these fragments are known, it is sometimes possible to deduce part or all of the original structure by taking advantage of overlapping pieces. For example, if a heptapeptide was composed of three glycines, two alanines, a leucine and a valine, many possible primary structures could be written. On the other hand, if partial hydrolysis gave two known tripeptide and two known dipeptide fragments, as shown on the right, simple analysis of the overlapping units identifies the original primary structure. Of course, this kind of structure determination is very inefficient and unreliable. First, we need to know the structures of all the overlapping fragments. Second, larger peptides would give complex mixtures which would have to be separated and painstakingly examined to find suitable pieces for overlapping. It should be noted, however, that modern mass spectrometry uses this overlap technique effectively. The difference is that bond cleavage is not achieved by hydrolysis, and computers assume the time consuming task of comparing a multitude of fragments.

5.7.2 N-Terminal Group Analysis

Over the years that chemists have been studying these important natural products, many techniques have been used to investigate their primary structure or amino acid sequence. Indeed, commercial instruments that automatically sequence peptides and proteins are now available. A few of the most important and commonly used techniques will be described here.

Identification of the N-terminal and C-terminal aminoacid units of a peptide chain provides helpful information. N-terminal analysis is accomplished by the Edman Degradation, which is outlined in the following diagram. A free amine function, usually in equilibrium with zwitterion species, is necessary for the initial bonding to the phenyl isothiocyanate reagent. The products of the Edman degradation are a thiohydantoin heterocycle incorporating the N-terminal amino acid together with a shortened peptide chain. Amine functions on a side-chain, as in lysine, may react with the isothiocyanate reagent, but do not give thiohydantoin products.

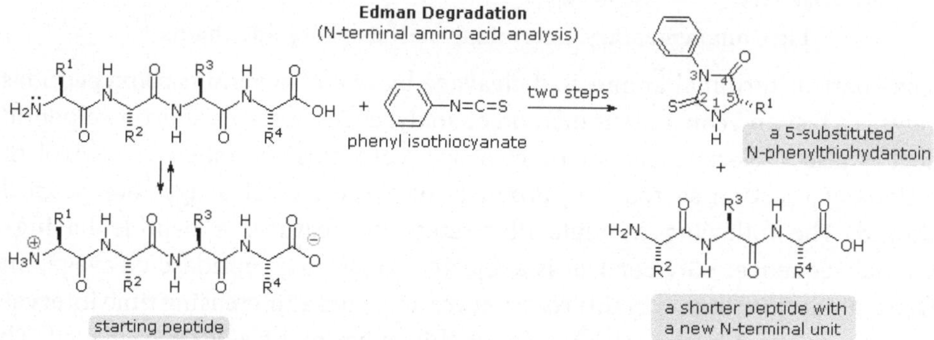

Fig Edman degradation

5.7.3 C-Terminal Group Analysis

Complementary C-terminal analysis of peptide chains may be accomplished chemically or enzymatically. The chemical analysis is slightly more complex than the Edman procedure. First, side-chain carboxyl groups and hydroxyl groups must be protected as amides or esters. Next, the C-terminal carboxyl group is activated as an anhydride and reacted with thiocyanate. The resulting acyl thiocyanate immediately cyclizes to a hydantoin ring, and this can be cleaved from the peptide chain in several ways, not described here. Depending on the nature of this final cleavage, the procedure can be modified to give a C-terminal acyl thiocyanate peptide product which automatically rearranges to a thiohydantoin incorporating the penultimate C-terminal unit. Thus, repetitive analyses may be conducted in much the same way they are with the Edman procedure.

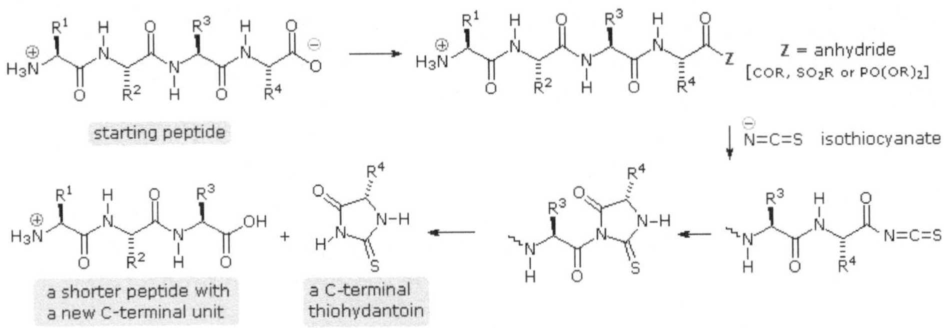

Fig Complementary C-terminal analysis of peptide chains

Enzymatic C-terminal amino acid cleavage by one of several carboxypeptidase enzymes is a fast and convenient method of analysis. Because the shortened peptide product is also subject to enzymatic cleavage, care must be taken to control the conditions of reaction so that the products of successive cleavages are properly monitored. The following example illustrates this feature. A peptide having a C-terminal sequence: ~Gly-Ser-Leu is subjected to carboxypeptidase cleavage, and the free aminoacids cleaved in this reaction are analyzed at increasing time intervals. By clicking on the diagram, the results of this experiment will be displayed. The leucine is cleaved first, the serine second, and the glycine third, as demonstrated by the sequential analysis. Of course, fourth and fifth units will also be released as time passes, but these products are not shown.

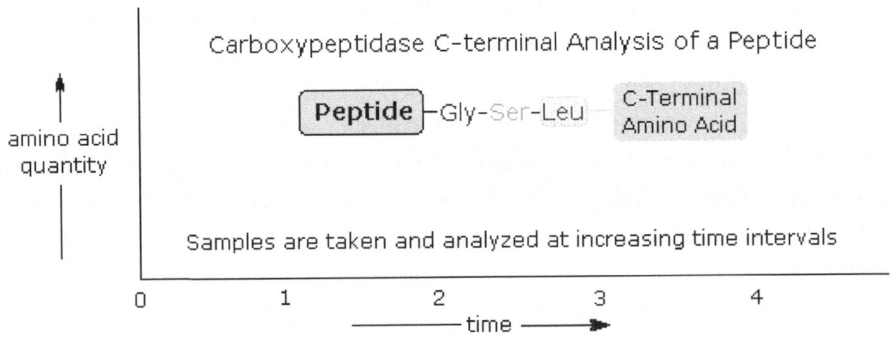

Fig Enzymatic C-terminal amino acid

5.7.4 Cyclic Peptides

If the carboxyl function at the C-terminus of a peptide forms a peptide bond with the N-terminal amine group a cyclic peptide is formed. Carboxyate and amine functions on side chains may also combine to form rings. Cyclic peptides are most commonly found in microorganisms, and often incorporate some D-amino acids as well as unusual amino acids such as ornithine (Orn). The decapeptide antibiotic gramacidin S, produced by a strain of Bacillus brevis, is one example of this interesting class of natural products. The structure of gramicidin S is shown in the following diagram. The atypical amino acids are colored. When using a shorthand notation for cyclic structures, the top line is written by the usual convention (N-group on the left), but vertical and lower lines must be adjusted to fit the bonding. Arrows on these bonds point in the CO-N direction of each peptide bond.

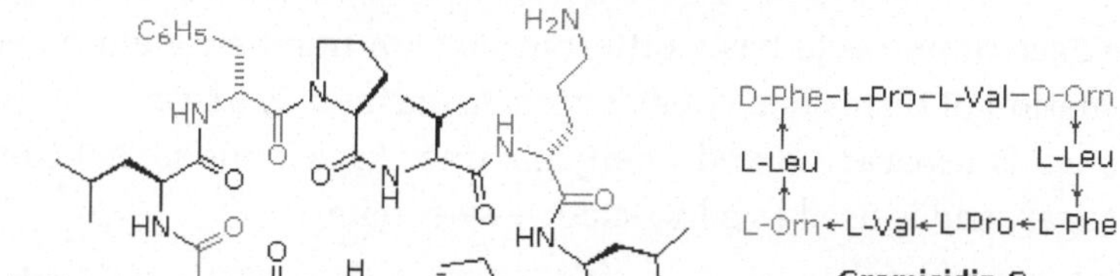

Gramicidin S
a peptide antibiotic

GLOSSARY

Acetaldehyde: Alcohol becomes acetaldehyde in the liver. It is then further oxidized into acetic acid and finally broken down into carbon dioxide for energy production. The acetaldehyde that is made in this process makes our faces red and triggers other effects when we drink alcohol.

Active Site of Enzyme: Enzymes speed up reactions in various substances. The part of the enzyme that is directly involved in this reaction is called the active site.

Aerobic Exercise: Aerobic exercise is endurance exercise that efficiently uses energy by the tricarboxylic acid (TCA) cycle while using large amounts of oxygen over an extended period. Running a marathon, swimming and tennis are examples of aerobics exercise.

Alanine: An amino acid that is an important source of energy for the liver.

Amine: Amines are often made in the body from amino acids. Many amine hormones influence the way our bodies function and how our nervous system works. Epinephrine (adrenaline), norepinephrine (noradrenalin), serotonin and histamine are all amines that are made in the body.

Amino Acid Infusion: Amino acid infusions were developed to help stomach and intestinal surgery patients get enough nutrition intravenously. These infusions have made it much easier to manage the nutrition of patients who are unable to take in enough food. Without these infusions, these patients would suffer from protein deficiency and severe malnourishment.

Arginine: An amino acid that works to ensure normal vascular function.

Aromatic Amino Acid: Amino acids that have an aromatic ring (phenylalanine, tryptophan and tyrosine) are called aromatic amino acids. These aromatic amino acids are used to make hormones and various amines in the body. They are also used as a component to make proteins.

Aspargine: An amino acid that was discovered and isolated from asparagus.

Aspartame: Each amino acid has a different taste. A number of amino acids have a combination of tastes like sweet-bitter. Aspartame is a sweetener made of two amino acids asparagine and phenylalanine. Aspartame is 200 times sweeter than sugar, and is used as a low-calorie sweetener.

Aspartic acid: An amino acid that is a quick-acting source of energy. Both asparagine and aspartic acid are positioned close to the tricarboxylic acid (TCA) cycle in energy production.

Basal Metabolism: Basal metabolism is the body's metabolization of energy at rest. Young people typically have a high basal metabolic rate but it slows as we get older.

BCAA Infusion: Branched-chain amino acids (BCAAs) help reduce the breakdown of proteins in the body. BCAA-rich infusions are widely used by hospitals to prevent the loss of protein in the body after surgery.

BCAA Oral Supplementation: Branched-chain amino acids (BCAAs) are often used to manage the nutrition of patients after surgery or for people with liver disease. BCAA oral supplementation is a treatment where patients take BCAA supplements orally.

Biodegradability: Biodegradation means the process of microorganisms that exist in nature gradually breaking down substances into natural elements. Amino acids are plentiful in nature and are quickly broken down and used in nature. This makes amino acids highly biodegradable and environmentally friendly.

Branched-Chain Amino Acid: Valine, isoleucine and leucine are called branched-chain amino acids (BCAAs). BCAAs are important for promoting protein growth in the body, improving liver function and providing energy during exercise.

Citric Acid: Citric acid is a substance in the tricarboxylic acid (TCA) cycle, which generates energy in the body. The TCA cycle starts with citric acid and is also called the "citric acid cycle."

Collagen: Collagen is a protein that makes up skin, ligaments and cartilage. Roughly one-third of the protein in the body is collagen. It is made from the amino acids alanine and proline.

Cuticle: The cuticle is the outermost part of the hair and is mainly made from protein. Daily environmental stresses such as getting permanent waves, applying hair dyes, exposure to ultraviolet radiation and washing can damage cuticles over time.

Cysteine: Cysteine is an amino acid that reduces the amount of black melanin pigmentation made in skin.

Cystine: Cystine is made when two cysteine amino acids join together. Cystine is plentiful in chicken meat and our bodies use cystine to make various substances including protein and glutathione. Cystine also promotes the normal function of immune cells.

Enzyme: Enzymes are proteins that speed up reactions to turn certain substances into other substances. Enzymes are a type of protein and are made from amino acids.

Epinephrine: Also called "adrenaline," epinephrine is a hormone that is secreted to help the body's immediate stress response. It does this by raising your heartbeat or blood pressure. Epinephrine is made in the body from the amino acids phenylalanine and tyrosine.

Essential amino acids: There are 20 amino acids that make up protein. Nine of them (histidine, isoleucine, leucine, lysine, methionine, phenylalanine, threonine, tryptophan and valine) cannot be made by our bodies and must be consumed in food. **These are called essential amino acids.**

Free Amino Acid: Some individual and unbound amino acids are contained in blood and cells, so that they can be immediately used by the body. These are called free amino acids, and are rapidly used in the body when they are needed.

Gluconeogenesis: Brain and muscle tissue continuously use glucose for energy and the body can make many different nutritious substances from sugars. Amino acids are typically used to make glucose. These processes are called gluconeogenesis.

Glucose: Glucose is one of the most important sources of energy in the body. In food our bodies get glucose by digesting starches. Blood sugar level is another way of referring to the glucose level in blood.

Glutamate: An amino acid that is plentiful in wheat and soy. Glutamate is the base of umami and free glutamates are found in kombu, tomatoes and cheese. It is one of the most usable amino acids for energy. Inside the body, glutamate is utilized as an important source of essential amino acids.

Glutamine: Glutamine is an amino acid that is needed to maintain normal gastrointestinal and muscle function.

Glutathione: Glutathione is a peptide (short chain of amino acids) made from glutamate, cysteine and glycine. Glutathione protects the body from oxidants and promotes the normal function of immune cells.

Glyceric Acid: Glyceric acid is an important substance for our bodies to make phospholipids, sugars and serine.

Glycine: A non-essential amino acid that is made in the body. Glycine is plentiful in the body. It acts as a transmitter in the central nervous system and facilitates the regulation of some physiological functions such as locomotion and sensory perception. Glycine makes up one-third of collagen.

High-Calorie Infusion: High-calorie infusions are infusions used for the nutritional management of patients who are unable to eat food orally. High-calorie infusions are used by combining a base solution (sugar and electrolyte infusion) and an amino acid infusion.

Histamine: Histamine is an important substance used for allergy response by the body. Histamine is made from histidine.

Histidine: An essential amino acid that is used to make histamine.

Immune Antibody: Immune antibodies are proteins that are made in order to attack and remove viruses and foreign invaders that enter the body. Antibodies are made from amino acids.

Immune Cells: Immune cells are a type of white blood cell. They are also called lymphocytes. There are various lymphocytes in the body and they work together to remove viruses and other substances that attack the body.

Isoleucine: A branched-chain amino acid (BCAA) together with valine and leucine. Isoleucine plays an important role in promoting protein growth in the body and providing energy during exercise.

Keratin: Keratin is a protein that is plentiful in hair, nails and skin and helps to protect the body. Keratin is rich in sulfur-containing amino acids such as methionine.

Koji: Koji or aspergillus oryzae is a culture that is used to make sake and soy sauce. Koji promotes the breakdown of protein and starch in grains and beans.

Lactic Acid: Our bodies make lactic acid when using sugars to produce energy. Our bodies only make a small amount of lactic acid in everyday life. However, if you suddenly start exercising or increase your exercise intensity, it can sharply increase the breakdown of sugars. This then causes a temporary and proportional increase the amount of lactic acid made. After the lactic acid is made, it is used by the body to produce energy, just like sugars.

Leucine: A branched-chain amino acid (BCAA) together with valine and isoleucine. Leucine plays an important role in promoting protein growth in the body and providing energy during exercise. Recent research indicates older persons need more leucine as they age.

Lysine: Lysine is one of the most commonly mentioned essential amino acids. Foods such as bread, wheat noodles and milled rice tend to be low in lysine.

Macrophage: Macrophages destroy viruses and other substances that attack the body. Microphages are always floating in the blood.

Melanin: Melanin pigments are the most important pigments in skin. Black melanin provides a black pigment. When you get sunburnt, the amount of black melanin increases. The amino acid cysteine reduces the amount of black melanin pigment made.

Methionine: An essential amino acid that is used to make many different substances needed in the body.

Natural Moisturizing Factor: Skin has a natural moisturizing factor (NMF) that is mainly made up of amino acids. The natural moisturizing factor retains moisture and keeps the skin smooth and supple by forming a layer of fat that is a barrier to the outside.

Non-Essential Amino Acid: Amino acids that can be made by our bodies are called non-essential amino acids. Other amino acids are essential amino acids. Scientists are increasingly discovering the many different functions of non-essential amino acids.

Norepinephrine: Also called noradrenaline, norepinephrine is a hormone that is secreted to help the body's immediate stress response by raising your blood pressure. Norepinephrine is made in the body from the amino acids phenylalanine and tyrosine.

Nucleic Acid: Nucleic acids, which include inosinic acid and guanylic acid, are known as umami substances. Inosinic acid is plentiful in meat and fish while guanylic acid is plentiful in mushrooms.

Oxaloacetic Acid: Oxaloacetic acid is a substance in the tricarboxylic acid (TCA) cycle that produces energy in the body.

Phenylalanine: An essential amino acid that is used to make many types of useful amines.

Phospholipid: Phospholipids are essential for making cell membranes. Phospholipids are abundant in cells and blood. Eggs and soy are foods with lots of phospholipids.

Porphyrin: The red pigment of blood comes from porphyrin in red blood cells. Porphyrin binds with iron to help iron bind with red blood cells.

Proline: An amino acid that is the main ingredient of collagen, which makes up skin. Proline is a quick-acting source of energy.

Protein: Proteins are made by chains of 10 or more amino acids or sometimes even hundreds of amino acids. Proteins make up many different components in the body including muscles and organs. Some hormones and enzymes are also made from proteins.

Pyrrolidone Carboxylic Acid: Pyrrolidone carboxylic acid (PCA) is an amino acid derivative. It is a moisturizing substance and is plentiful in the natural moisturizing factor (NMF) of skin.

Pyruvic Acid: Alanine and sugars are first converted to pyruvic acid. This is then converted to citric acid before entering the tricarboxylic acid (TCA) cycle where it is burned for energy. Pyruvic acid is one of the inputs of the TCA cycle.

Sarcopenia: Sarcopenia is the natural loss of skeletal muscle mass as we get older. As sarcopenia progresses, our strength and endurance typically decreases. It can even lead to the need for assistive care.

Serine: An amino acid used to make phospholipids and glyceric acid.

Serotonin: Serotonin in the brain helps to regulate body temperature, sleep and appetite. When animals repeatedly exercise for long periods of time the amount of serotonin increases in the brain. That's why scientists believe there is a strong connection between serotonin and fatigue. Serotonin is made in the body from the amino acid tryptophan.

Sulfur-Containing Amino Acid: Amino acids such as methionine and cystine are called "sulfur-containing amino acids" because they contain sulfur.

Theanine: Theanine is an amino acid that was discovered in tea leaves and is plentiful in high-grade teas such as green tea. Theanine is metabolized into glutamate and ethylamine in the body.

Threonine: An essential amino acid that is used to make the active site of enzymes.

Tricarboxylic Acid Cycle: The tricarboxylic acid (TCA) cycle, also called the citric acid cycle, breaks down amino acids, sugars and fats to produce energy.

Tryptophan: An essential amino acid used to make many types of useful amines.

Turnover: Turnover is the process of protein in the body being replaced with new protein. This can take several minutes or even several months. Skin turnover, for example, is the cycle of new skin being made, used on the outermost part of the body and shed off when it is used up.

Tyrosine: Tyrosine is used to make many types of useful amines. Tyrosine is an aromatic amino acid together with phenylalanine and tryptophan.

Umami Substance: A substance that produces umami, one of the five basic tastes. Known umami substances include glutamate from kombu, inosinic acid from bonito fish shavings, and guanylic acid from dried shiitake mushrooms. Glutamate is plentiful in cheeses and vegetables. Inosinic acid is plentiful in meat and fish. Guanylic acid is plentiful in mushrooms.

Valine: A branched-chain amino acid (BCAA) together with isoleucine and leucine. Valine plays an important role in promoting protein growth in the body and providing energy during exercise.

CHAPTER- 6

PROTEINS

6.1 INTRODUCTION

Proteins are any of a group of complex organic compounds containing carbon, hydrogen, nitrogen and sulfur. Proteins, the principle constituents of the protoplasm of all cells, are of high molecular weight and consist of alpha-amino acids joined by peptide linkages. Different amino acids are commonly found in proteins, each protein having a unique, genetically defined amino-acid sequence, which determines its specific shape and function. They serve as enzymes, structural elements, hormones, immunoglobulins, etc. and are involved in oxygen transport, muscle contraction, electron transport, and other activities.

The importance of adequate protein intake to proper immune function has been extensively studied. The most severe effects of "Protein-Calorie Malnutrition" (PCM) are on cell-mediated immunity, although all facets of immune function are ultimately affected. PCM is not, however, a single nutrient deficiency. It is normally associated with multiple nutrient deficiencies, and some immune dysfunctions attributed to PCM are most likely due to these other factors. Partial deficiencies of dietary vitamins produce a comparatively greater depression on immune functions than do partial protein deficiencies. Non-the less, adequate protein is essential for optimal immune function.

High protein diets are not recommended for individuals with kidney or liver disease. There are no other known side effects. A macromolecule made from amino acids; some consist of one, long polypeptide chain while others are made from several intertwined polypeptides (proteios = first place)

6.2 FEATURES

Proteins are about 50% of the dry weight of most cells, and are the most structurally complex macromolecules known. Each type of protein has its own unique structure and function.

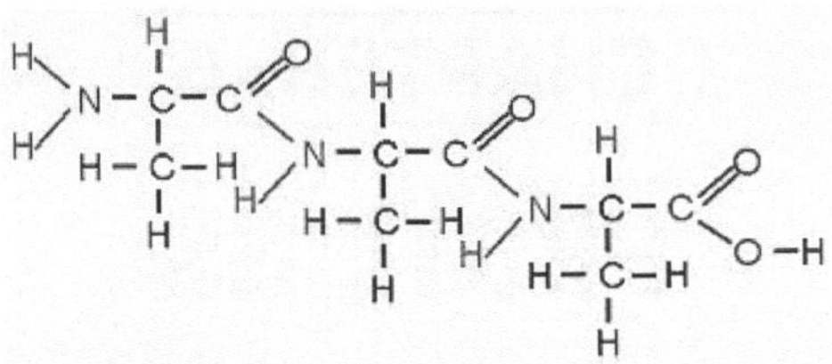

Fig Protein macromolecules

6.3 PROTEINS AS POLYMER

A large molecule made of many, smaller, repeating, identical or similar subunits (poly = many; mer = a part) Polymers are any kind of large molecules made of repeating identical or similar subunits called monomers. Proteins are polymers of about 20 amino acids (the monomer).

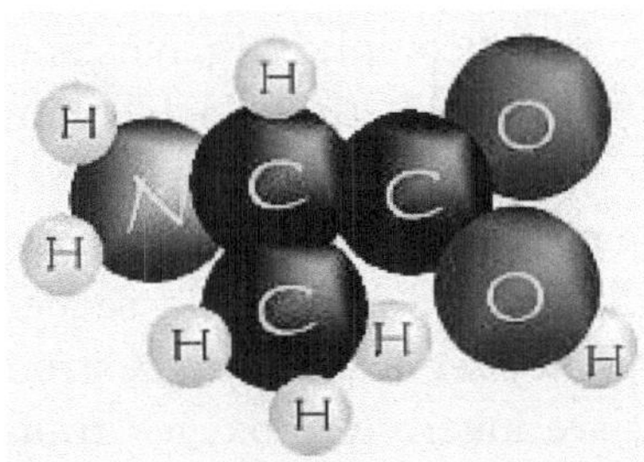

Amino acids are built from a central carbon bonded to four different groups.

- ☆ Hydrogen (–H),
- ☆ Amino group (–NH2),
- ☆ Carboxyl group (–COOH),

and some side chain symbolized by "R". In the approximately-20 amino acids found in our bodies, what varies is the side chain. Some side chains are hydrophilic while others are hydrophobic. Since these side chains stick out from the backbone of the molecule, they help determine the properties of the protein made from them.

Most naturally-occurring amino acids are the l- form, whereas synthetically-produced amino acids give a 50:50 mixture. Notice that these molecules are mirror images of each other, thus there is no way you can rotate one molecule to make it look like the other.

Fig naturally-occurring amino acids

To form protein, the amino acids are linked by dehydration synthesis to form peptide bonds. The chain of amino acids is also known as a polypeptide. Some proteins contain only one polypeptide chain while others, such as hemoglobin, contain several polypeptide chains all twisted together. The sequence of amino acids in each polypeptide or protein is unique to that protein, so each protein has its own, unique 3-D shape or native conformation. If even one amino acid in the sequence is changed, that can potentially change the protein's ability to function. For example, sickle cell anemia is caused by a change in only one nucleotide in the DNA sequence that causes just one amino acid in one of the hemoglobin polypeptide molecules to be different. Because of this, the whole red blood cell ends up being deformed and unable to carry oxygen properly.

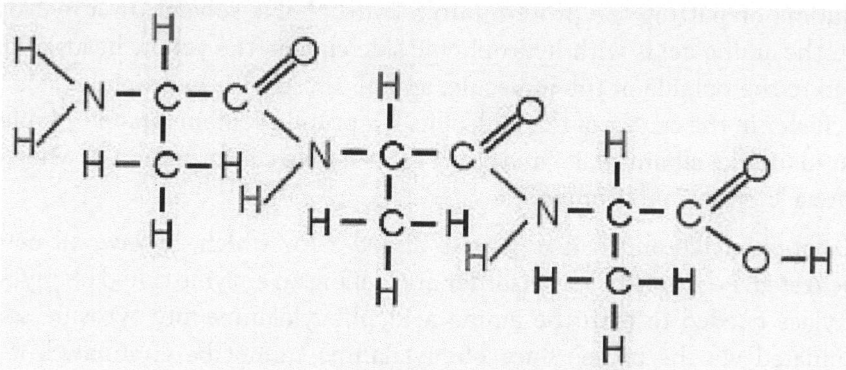

Fig polypeptide chain formation

As a polypeptide chain forms, it naturally twists and bends into its native conformation. One of the things that helps determine the native conformation of a protein is the side chains of all the amino acids involved. Remember some amino acid side chains are hydrophobic while others are hydrophilic. In this case, the "likes" attract: all the hydrophobic side chains (here represented by yellow beads) try to "get together" in the center of the molecule, away from the watery environment, while the hydrophilic side chains are attracted to the outside of the molecule, near the watery environment.

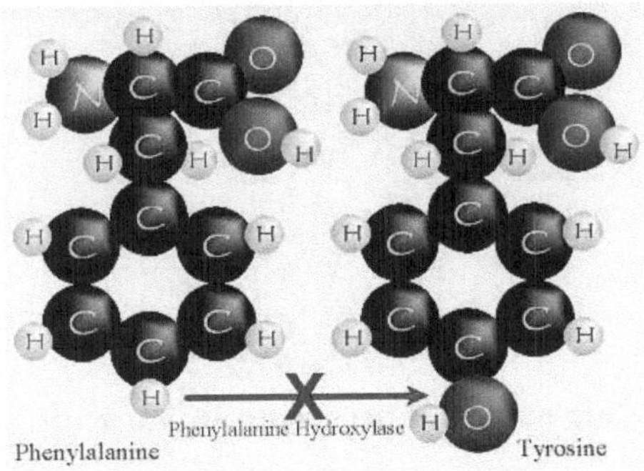

Additionally, some of the hydrophilic side chains have groups of atoms attached that make them acidic, while others have groups attached that make them basic. Side chains with acidic ends are attracted to side chains with basic ends, and can form ionic bonds. Thus, the side chains interacting with each other help to hold the protein in its native conformation.

Denaturation is when a protein loses its native conformation. For example, egg white (also called albumen) contains a protein called albumin which is water-soluble. However, if heated, albumin becomes denatured and loses its ability to be water-soluble. There are several possible things that can denature proteins including changing the temperature (adding heat), changing the pH or salt concentration of the solution, or putting the protein into a hydrophobic solvent. In a hydrophobic solvent, the amino acids with hydrophobic side chains (the yellow beads) would all try to go to the outside of the molecule, and all those with hydrophilic side chains would cluster in the center of the molecule. If a protein remains water-soluble when denatured (unlike albumin), it can return to its native conformation if/when placed back into a "normal" environment.

PKU or phenylketonuria is a genetic disorder for which, by law, all newborns must be tested. People with this disorder are lacking an enzyme called phenylalanine hydroxylase needed to turn the amino acid, phenylalanine into tyrosine so it can be eliminated via the urine. Since phenylalanine cannot be eliminated, it builds up and in children whose nervous systems are still developing, causes mental retardation. If PKU is detected shortly after birth, the child is put on a special diet low in phenylalanine and has normal mental development. Since phenylalanine is an amino acid that is needed by the body for normal growth and development, the goal is to limit intake to just what the child needs without any excess that could build up in his/her system.

6.4 STRUCTURE OF PROTEINS

Now we see how amino acids combine to make proteins and what is meant by the primary, secondary and tertiary structures of proteins. It only applies to proteins consisting of more than one polypeptide chain. If you were to draw the structure of a general 2-amino acid, you would probably draw it like this:

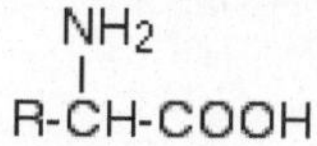

However, for drawing the structures of proteins, we usually twist it so that the "R" group sticks out at the side. It is much easier to see what is happening if you do that.

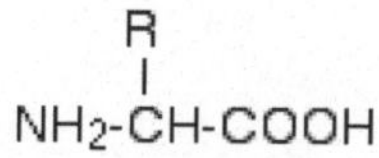

That means that the two simplest amino acids, glycine and alanine, would be shown as:

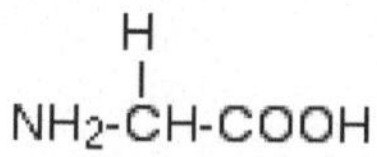

glycine

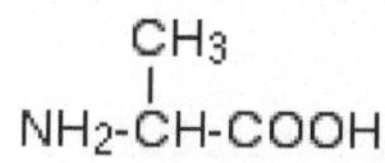

alanine

6.4.1 Primary Structure

At its simplest, the term is used to describe the order of the amino acids joined together to make the protein. In other words, if you replaced the "R" groups in the last diagram by real groups you would have the primary structure of a particular protein.

This primary structure is usually shown using abbreviations for the amino acid residues. These abbreviations commonly consist of three letters or one letter.

If you followed the protein chain all the way to its left-hand end, you would find an amino acid residue with an unattached -NH2 group. The N-terminal is always written on the left of a diagram for a protein's primary structure - whether you draw it in full or use these abbreviations.

The wider definition of primary structure includes all the features of a protein which are a result of covalent bonds. Obviously, all the peptide links are made of covalent bonds, so that isn't a problem. But there is an additional feature in proteins which is also covalently bound. It involves the amino acid cysteine.

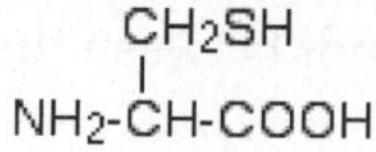

cysteine

If two cysteine side chains end up next to each other because of folding in the peptide chain, they can react to form a sulphur bridge. This is another covalent link and so some people count it as a part of the primary structure of the protein.

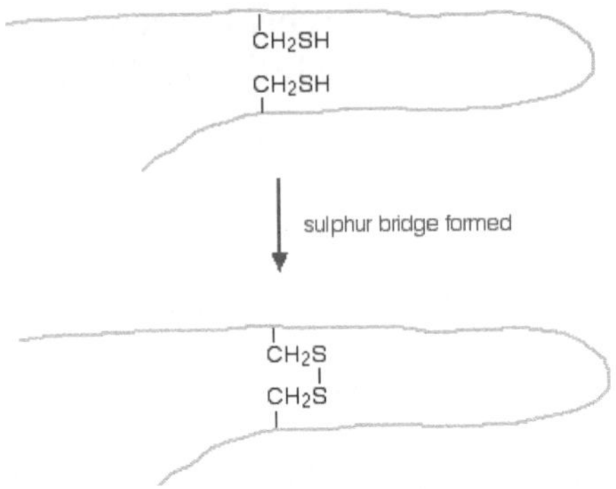

Because of the way sulphur bridges affect the way the protein folds, other people count this as a part of the tertiary structure. This is obviously a potential source of confusion!

6.4.2 secondary structure

Within the long protein chains there are regions in which the chains are organised into regular structures known as alpha-helices (alpha-helixes) and beta-pleated sheets. These are the secondary structures in proteins.

These secondary structures are held together by hydrogen bonds. These form as shown in the diagram between one of the lone pairs on an oxygen atom and the hydrogen attached to a nitrogen atom:

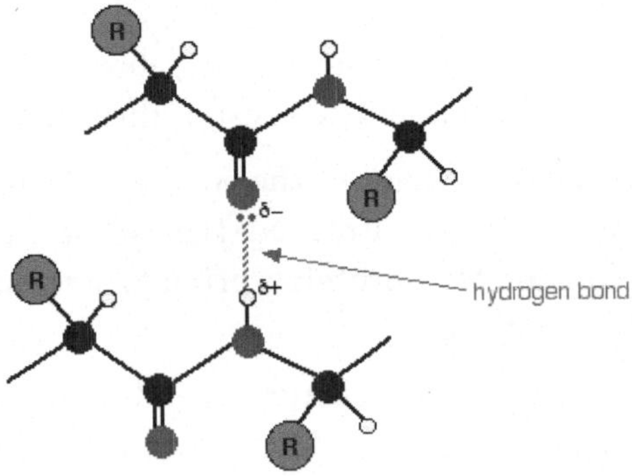

You must also find out exactly how much detail you need to know about this next bit. It may well be that all you need is to have heard of an alpha-helix and know that it is held together by hydrogen bonds between the C=O and N-H groups.

The alpha-helix

In an alpha-helix, the protein chain is coiled like a loosely-coiled spring. The "alpha" means that if you look down the length of the spring, the coiling is happening in a clockwise direction as it goes away from you. The next diagram shows how the alpha-helix is held together by hydrogen bonds. This is a very simplified diagram, missing out lots of atoms. We'll talk it through in some detail after you have had a look at it.

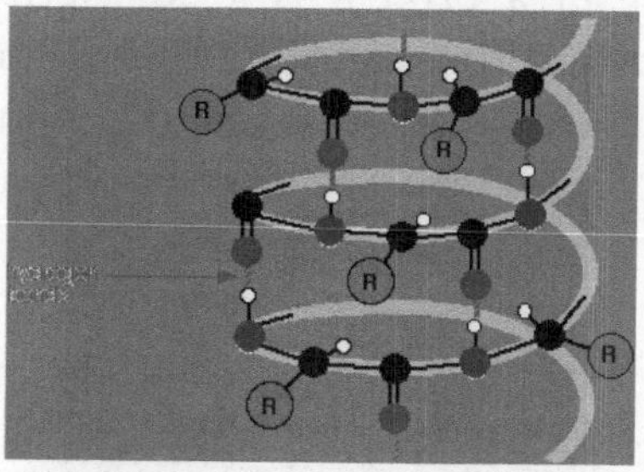

If you had a whole number of amino acid residues per turn, each group would have an identical group underneath it on the turn below. Hydrogen bonding can't happen under those circumstances. Each turn has 3 complete amino acid residues and two atoms from the next one. That means that each turn is offset from the ones above and below, such that the N-H and C=O groups are brought into line with each other.

Beta-pleated sheets

In a beta-pleated sheet, the chains are folded so that they lie alongside each other. The next diagram shows what is known as an "anti-parallel" sheet. All that means is that next-door chains are heading in opposite directions. Given the way this particular folding happens, that would seem to be inevitable.

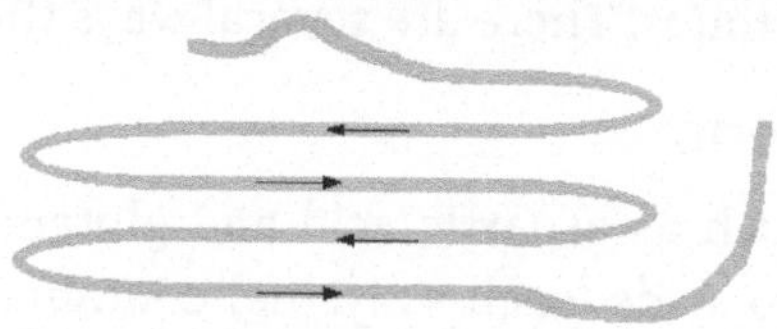

The folded chains are again held together by hydrogen bonds involving exactly the same groups as in the alpha-helix.

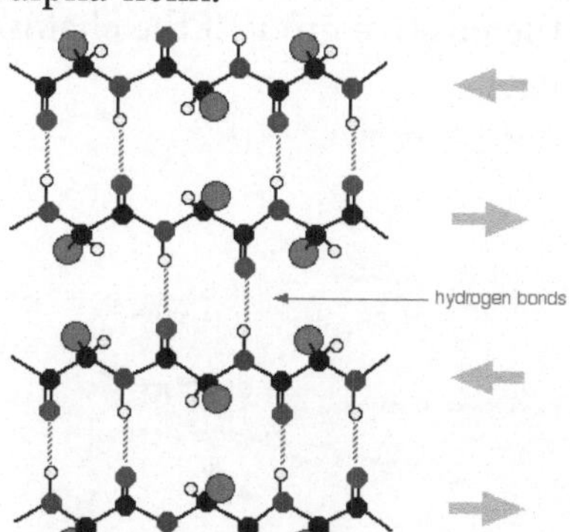

6.4.3 Tertiary structure

The tertiary structure of a protein is a description of the way the whole chain (including the secondary structures) folds itself into its final 3-dimensional shape. This is often simplified into models like the following one for the enzyme dihydrofolate reductase. Enzymes are, of course, based on proteins.

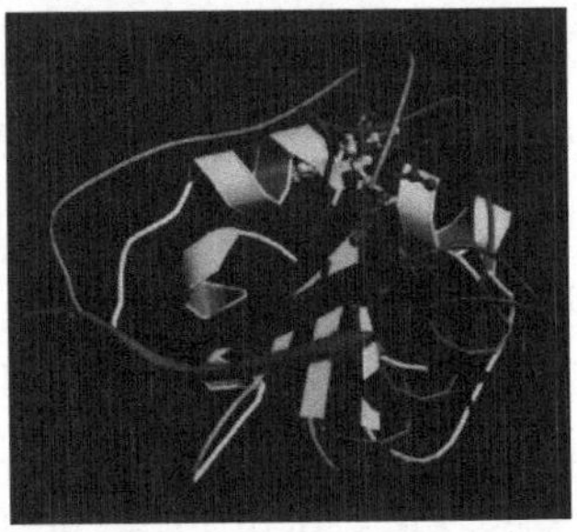

The model shows the alpha-helices in the secondary structure as coils of "ribbon". The beta-pleated sheets are shown as flat bits of ribbon ending in an arrow head. The bits of the protein chain which are just random coils and loops are shown as bits of "string". The colour coding in the model helps you to track your way around the structure - going through the spectrum from dark blue to end up at red. You will also notice that this particular model has two other molecules locked into it. These are the two molecules whose reaction this enzyme catalyses.

The tertiary structure of a protein is held together by interactions between the the side chains - the "R" groups. There are several ways this can happen.

6.4.3.1 Ionic interactions

Some amino acids (such as aspartic acid and glutamic acid) contain an extra -COOH group. Some amino acids (such as lysine) contain an extra -NH2 group. You can get a transfer of a hydrogen ion from the -COOH to the -NH2 group to form zwitterions just as in simple amino acids. You could obviously get an ionic bond between the negative and the positive group if the chains folded in such a way that they were close to each other.

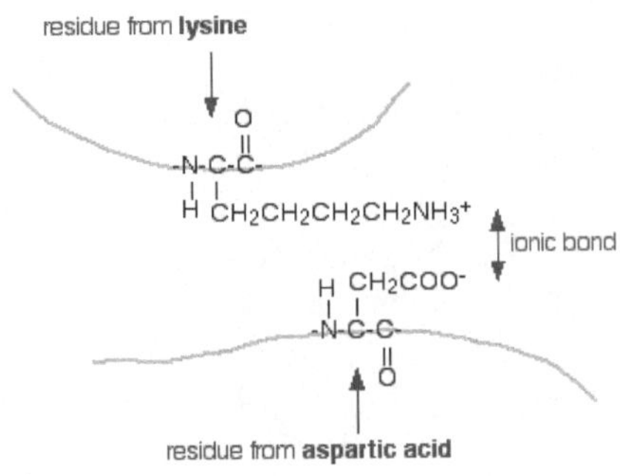

6.4.3.2 Hydrogen bonds

Lots of amino acids contain groups in the side chains which have a hydrogen atom attached to either an oxygen or a nitrogen atom. This is a classic situation where hydrogen bonding can occur. For example, the amino acid serine contains an -OH group in the side chain. You could have a hydrogen bond set up between two serine residues in different parts of a folded chain.

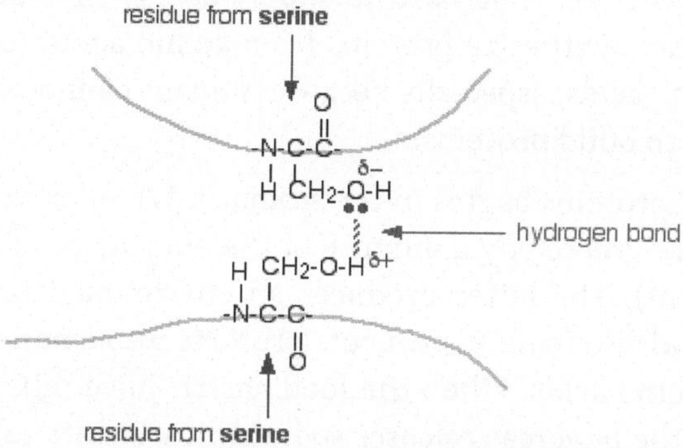

You could easily imagine similar hydrogen bonding involving -OH groups, or -COOH groups, or -CONH2 groups, or -NH2 groups in various combinations - although you would have to be careful to remember that a -COOH group and an -NH2 group would form a zwitterion and produce stronger ionic bonding instead of hydrogen bonds.

Several amino acids have quite large hydrocarbon groups in their side chains. A few examples are shown below. Temporary fluctuating dipoles in one of these groups could induce opposite dipoles in another group on a nearby folded chain. The dispersion forces set up would be enough to hold the folded structure together.

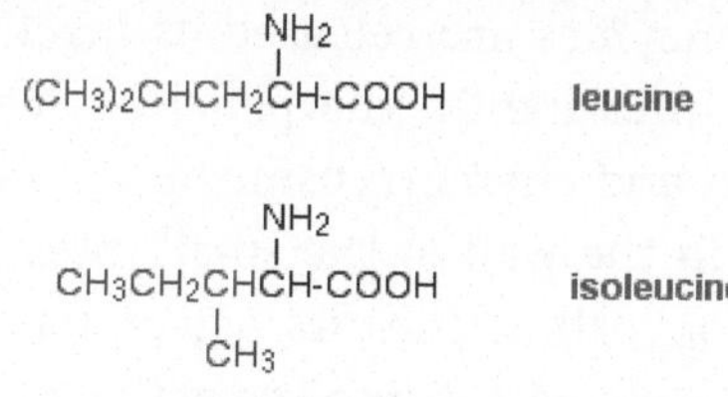

6.5. PROTEIN METABOLISM

Much of the body is made of protein, and these proteins take on a myriad of forms. They represent cell signaling receptors, signaling molecules, structural members, enzymes, intracellular trafficking components, extracellular matrix scaffolds, ion pumps, ion channels, oxygen and CO2 transporters (hemoglobin). That is not even the complete list! There is protein in bones (collagen), muscles, and tendons; the hemoglobin that transports oxygen; and enzymes that catalyze all biochemical reactions. Protein is also used for growth and repair. Amid all these necessary functions, proteins also hold the potential to serve as a metabolic fuel source. Proteins are not stored for later use, so excess proteins must be converted into glucose or triglycerides, and used to supply energy or build energy reserves. Although the body can synthesize proteins from amino acids, food is an important source of those amino acids, especially because humans cannot synthesize all of the 20 amino acids used to build proteins.

The digestion of proteins begins in the stomach. When protein-rich foods enter the stomach, they are greeted by a mixture of the enzyme pepsin and hydrochloric acid (HCl; 0.5 percent). The latter produces an environmental pH of 1.5–3.5 that denatures proteins within food. Pepsin cuts proteins into smaller polypeptides and their constituent amino acids. When the food-gastric juice mixture (chyme) enters the small intestine, the pancreas releases sodium bicarbonate to neutralize the HCl. This helps to protect the lining of the intestine. The small intestine also releases digestive hormones, including secretin and CCK, which stimulate digestive processes to break down the proteins further. Secretin also stimulates the pancreas to release sodium bicarbonate. The pancreas releases most of the digestive enzymes, including the proteases trypsin, chymotrypsin, and elastase, which aid protein digestion. Together, all of these enzymes break complex proteins into smaller individual amino acids, which are then transported across the intestinal mucosa to be used to create new proteins, or to be converted into fats or acetyl CoA and used in the Krebs cycle.

In order to avoid breaking down the proteins that make up the pancreas and small intestine, pancreatic enzymes are released as inactive proenzymes that are only activated in the small intestine. In the pancreas, vesicles store trypsin and chymotrypsin as trypsinogen and chymotrypsinogen. Once released into the small intestine, an enzyme found in the wall of the small intestine, called enterokinase, binds to trypsinogen and converts it into its active form, trypsin. Trypsin then binds to chymotrypsinogen to convert it into the active chymotrypsin. Trypsin and chymotrypsin break down large proteins into smaller peptides, a process called proteolysis. These smaller peptides are catabolized into their constituent amino acids, which are transported across the apical surface of the intestinal mucosa in a process that is mediated by sodium-amino acid transporters. These transporters bind sodium and then bind the amino acid to transport it across the membrane. At the basal surface of the mucosal cells, the sodium and amino acid are released. The sodium can be reused in the transporter, whereas the amino acids are transferred into the bloodstream to be transported to the liver and cells throughout the body for protein synthesis.

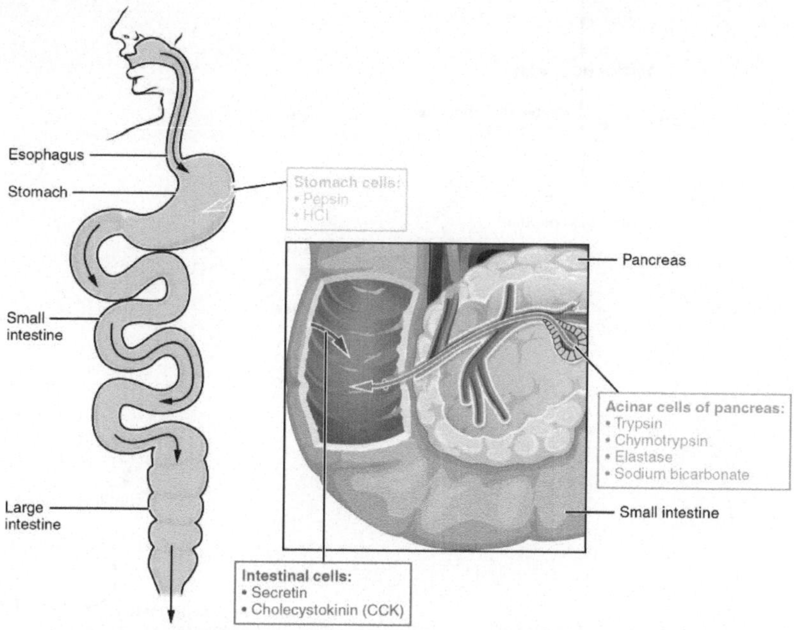

Fig breaking of complex proteins into smaller amino acids

Freely available amino acids are used to create proteins. If amino acids exist in excess, the body has no capacity or mechanism for their storage; thus, they are converted into glucose or ketones, or they are decomposed. Amino acid decomposition results in hydrocarbons and nitrogenous waste. However, high concentrations of nitrogen are toxic. The urea cycle processes nitrogen and facilitates its excretion from the body.

6.5.1 Urea Cycle

The urea cycle is a set of biochemical reactions that produces urea from ammonium ions in order to prevent a toxic level of ammonium in the body. It occurs primarily in the liver and, to a lesser extent, in the kidney. Prior to the urea cycle, ammonium ions are produced from the breakdown of amino acids. In these reactions, an amine group, or ammonium ion, from the amino acid is exchanged with a keto group on another molecule. This transamination event creates a molecule that is necessary for the Krebs cycle and an ammonium ion that enters into the urea cycle to be eliminated. In the urea cycle, ammonium is combined with CO2, resulting in urea and water. The urea is eliminated through the kidneys in the urine.

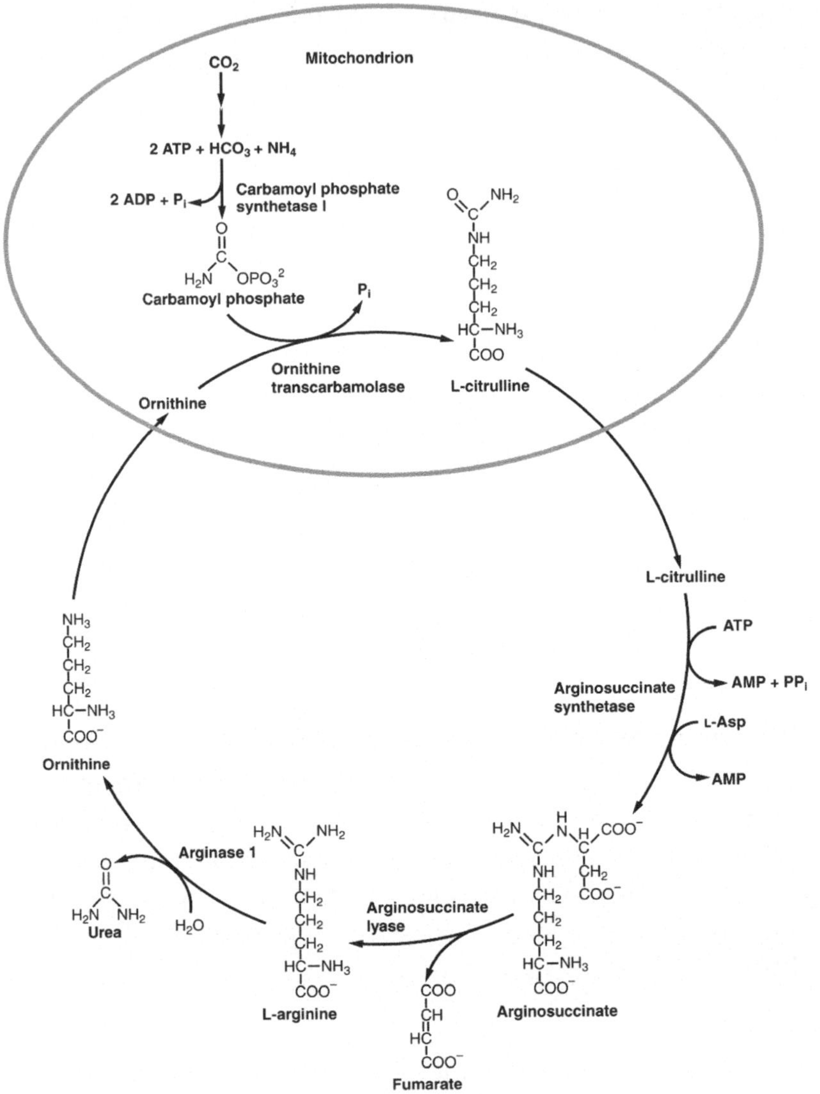

Fig. Urea Cycle

6.5.2 Metabolism

Pyruvate Dehydrogenase Complex Deficiency and Phenylketonuria

Pyruvate dehydrogenase complex deficiency (PDCD) and phenylketonuria (PKU) are genetic disorders. Pyruvate dehydrogenase is the enzyme that converts pyruvate into acetyl CoA, the molecule necessary to begin the Krebs cycle to produce ATP. With low levels of the pyruvate dehydrogenase complex (PDC), the rate of cycling through the Krebs cycle is dramatically reduced. This results in a decrease in the total amount of energy that is produced by the cells of the body. PDC deficiency results in a neurodegenerative disease that ranges in severity, depending on the levels of the PDC enzyme. It may cause developmental defects, muscle spasms, and death. Treatments can include diet modification, vitamin supplementation, and gene therapy; however, damage to the central nervous system usually cannot be reversed.

6.6 BIOLOGICAL FUNCTIONS OF PROTEINS

- ☆ **Membrane Proteins:** Proteins and lipids form the major structural components of cell membrane. The membrane associated proteins consists of intrinsic proteins and extrinsic proteins. Many enzymes and enzyme systems are associated with membrane proteins i.e. components of electron transport system.
- ☆ **Enzymes:** Enzymes are proteins produced within an organism which are capable of catalyzing specific catalytic reactions. They are biocatalysts which influence the rate f a chemical reaction, usually without undergoing any change themselves.
- ☆ **Hormones:** Several hormones are peptides and proteins. They play an important role in the regulation of metabolic reactions.
- ☆ **Blood proteins:** The blood proteins include plasma proteins and haemoglobin.
    - Plasma proteins are mainly albumin, globulins and fibrinogen. Albumin maintains the colloid osmotic pressure of plasma and transport materials. Globulins inhibit proteolytic enzymes, and helps to immunosystem of the body. Fibrinogen helps in blood clotting. Lipoproteins are concerned with transport of fat in blood.
    - Haemoglobin is a conjugated protein consisting ofglobin and haeme. Haemoglobin carries oxygen in the form of oxyhaemoglobin in blood.
- ☆ Antibodies: Some antibodies like gramicidin S, tyrocidin and penicillin G are peptides.
- ☆ **Nucleoproteins:** These are conjugated proteins of cell nuclei.
- ☆ **Multiple protein Assemblies:** These contain several compounds which together from a functional unit.
    - **Collagen:** A fibrous proteins found in many connective tissue.
    - **Flagella:** Microtubules of flagella are made up of tubules.
    - **Actionmyosm complex of muscle:** Muscle fibres consists of thin filaments of action, tropomyosin and troponin which make up the 1 or light bands and thick filaments of myosin which make up the A or dark bands.

GLOSSARY

Chymotrypsin: pancreatic enzyme that digests protein

Chymotrypsinogen: proenzyme that is activated by trypsin into chymotrypsin

Elastase: pancreatic enzyme that digests protein

Enterokinase: enzyme located in the wall of the small intestine that activates trypsin

Inactive proenzymes: forms in which proteases are stored and released to prevent the inappropriate digestion of the native proteins of the stomach, pancreas, and small intestine

pepsin: enzyme that begins to break down proteins in the stomach

proteolysis: process of breaking proteins into smaller peptides

secretin: hormone released in the small intestine to aid in digestion

sodium bicarbonate: anion released into the small intestine to neutralize the pH of the food from the stomach

transamination: transfer of an amine group from one molecule to another as a way to turn nitrogen waste into ammonia so that it can enter the urea cycle

trypsin: pancreatic enzyme that activates chymotrypsin and digests protein

trypsinogen: proenzyme form of trypsin

urea cycle: process that converts potentially toxic nitrogen waste into urea that can be eliminated through the kidneys

CHAPTER-7

LIPIDS AND FATTY ACID

7.1 FAT (LIPIDS)

Lipids are a broad group of diverse and naturally occurring, organic compounds. Lipids play a large role when it comes to energy storage and the structure of cells. They are divided into four main groups consisting of fats, lipids, hormones and steroids. The difference between fats and lipids is that fats are a subset of lipids. Lipids is a broad category that describes a group of about 7,000 fat-soluble, hydrophobic (not soluble in water) and naturally occurring organic compounds, found in animal or plant tissues. Lipids include compounds such as fatty acids, cholesterol, lipoproteins, phospholipids, steroids, waxes and even fat soluble vitamins. One of the subcategories of lipids, the fatty acids, is crucial to understand the differences between fats and oils.

Fats, like protein and carbohydrates, are a nutrient group that supplies energy to the body. Fats are organic compounds that come in liquid or solid form. They are composed of saturated and unsaturated fatty acids. Saturated fat causes high levels of bad cholesterol, known as LDL, and can be found in animal products and vegetable oils. Unsaturated fats help to lower bad cholesterol, but still must be consumed in moderation because of their caloric content. Unsaturated fats are found in most liquid vegetable oils.

Fats refer to lipids that are solid at room temperature (have a high melting point). They are solid, mainly because they have a higher proportion of saturated or trans fatty acids and their hydrocarbon chains are long. For more details on fats' melting points, read here.

Oils, in a nutritional sense, refer to lipids that are liquid at room temperature (have a low melting point). Similar to fats, oils mainly consist of triglycerides. However, unlike fats, the predominant fatty acids in those triglycerides are unsaturated, which make them liquid at room temperature. Although some oils can be non-edible, such as petrochemical products, in the nutritional context, they refer to edible oils that are mainly of plant origin, but also of animal origin.

Cholesterol, oils, butter, and margarine are all fats, or lipids, but each has slightly different characteristics. The one common attribute shared by lipids is that they are soluble in organic solvents but not soluble in water. The term fat is usually applied to lipids that are solid at room temperature, and the term oil is applied to lipids that are liquid at room temperature. The most commonly consumed form of lipid is triglyceride, which consists of three fatty acids and one glycerol molecule. Despite the numerous forms of lipids, we can obtain them all from the food supply, and we are also capable of making many types of lipids by combining carbon units from other substances. Nearly every cell in the body has the capacity to make cholesterol, which is why a person can have a high blood cholesterol leveleven when on a low-cholesterol diet. We can also manufacture phospholipids, triglycerides, and oils. In fact, it is this ability to effectively manufacture different types of lipids that limits the necessity to consume large amounts of lipids.

7.2 STRUCTURE AND FUNCTION

7.2.1 Fats

A certain amount of fat, between 20 and 35 percent of total consumed calories, is necessary to ensure a sufficient energy and nutrient intake. The fat-soluble vitamins-vitamins A,D,E, and K-must be delivered in a fat package. The essential fatty acids, which are needed for specific body functions but that we are incapable of synthesizing, must also be consumed. Some dietary fat is also needed to give us a feeling of satiety during the meal, creating the important physiological signal that it is time to stop eating.

The "tail" of a fatty acid is a long hydrocarbon chain, making it hydrophobic. The "head" of the molecule is a carboxyl group which is hydrophilic. Fatty acids are the main component of soap, where their tails are soluble in oily dirt and their heads are soluble in water to emulsify and wash away the oily dirt. However, when the head end is attached to glycerol to form a fat, that whole molecule is hydrophobic.

The terms saturated, mono-unsaturated, and poly-unsaturated refer to the number of hydrogens attached to the hydrocarbon tails of the fatty acids as compared to the number of double bonds between carbon atoms in the tail. Fats, which are mostly from animal sources, have all single bonds between the carbons in their fatty acid tails, thus all the carbons are also bonded to the maximum number of hydrogen's possible. Since the fatty acids in these triglycerides contain the maximum possible amount of hydrogen's, these would be called saturated fats. The hydrocarbon chains in these fatty acids are, thus, fairly straight and can pack closely together, making these fats solid at room temperature.

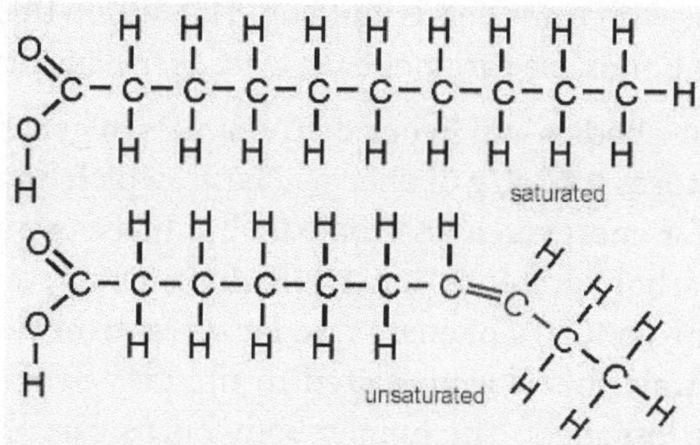

Oils, mostly from plant sources, have some double bonds between some of the carbons in the hydrocarbon tail, causing bends or "kinks" in the shape of the molecules. Because some of the carbons share double bonds, they're not bonded to as many hydrogen's as they could if they weren't double bonded to each other. Therefore these oils are called unsaturated fats.

Many people have heard that the unsaturated fats are "healthier" than the saturated ones. Hydrogenated vegetable oil (as in shortening and commercial peanut butters where a solid consistency is sought) started out as "good" unsaturated oil. However, this commercial product has had all the double bonds artificially broken and hydrogen's artificially added (in a chemistry lab-type setting) to turn it into saturated fat that bears no resemblance to the original oil from which it came (so it will be solid at room temperature).

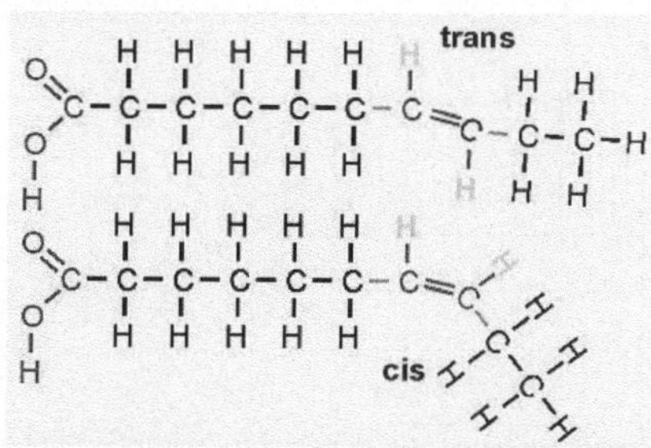

In unsaturated fatty acids, there are two ways the pieces of the hydrocarbon tail can be arranged around a C=C double bond. In cis bonds, the two pieces of the carbon chain on either side of the double bond are either both "up" or both "down," such that both are on the same side of the molecule. In trans bonds, the two pieces of the molecule are on opposite sides of the double bond, that is, one "up" and one "down" across from each other. Naturally-occurring unsaturated vegetable oils have almost all cis bonds, but using oil for frying causes some of the cis bonds to convert to trans bonds. If oil is used only once like when you fry an egg, only a few of the bonds do this so it's not too bad. However, if oil is constantly reused, like in fast food French fry machines, more and more of the cis bonds are changed to trans until significant numbers of fatty acids with trans bonds build up. The reason this is of concern is that fatty acids with trans bonds are carcinogenic, or cancer-causing.

We need fats in our bodies and in our diet. Animals in general use fat for energy storage because fat stores 9 KCal/g of energy. Plants, which don't move around, can afford to store food for energy in a less compact but more easily accessible form, so they use starch (a carbohydrate, NOT A LIPID) for energy storage. Carbohydrates and proteins store only 4 KCal/g of energy, so fat stores over twice as much energy/gram as fat. By the way, this is also related to the idea behind some of the high-carbohydrate weight loss diets. The human body burns carbohydrates and fats for fuel in a given proportion to each other. The theory behind these diets is that if they supply carbohydrates but not fats, then it is hoped that the fat needed to balance with the sugar will be taken from the dieter's body stores. Fat is also is used in our bodies to a) cushion vital organs like the kidneys and b) serve as insulation, especially just beneath the skin.

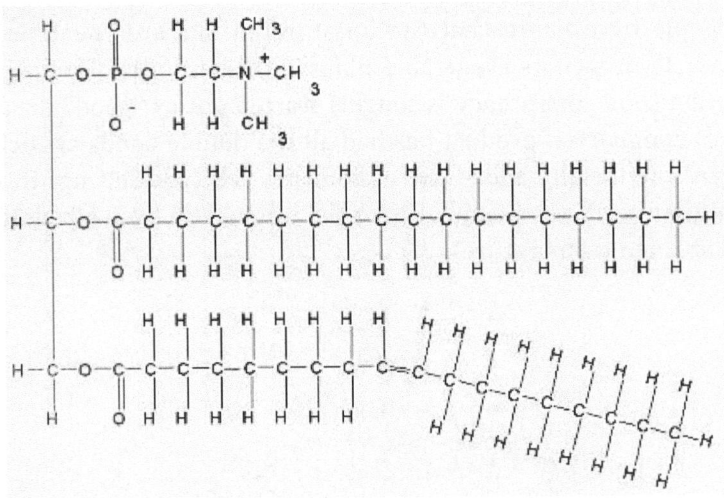

7.2.2 Lipids

All Lipids are hydrophobic: that's the one property they have in common. This group of molecules includes fats and oils, waxes, phospholipids, steroids (like cholesterol), and some other related compounds.

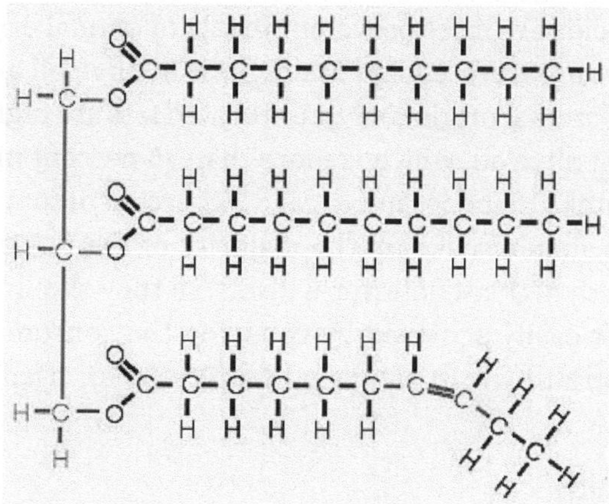

Fats and oils are made from two kinds of molecules: glycerol (a type of alcohol with a hydroxyl group on each of its three carbons) and three fatty acids joined by dehydration synthesis. Since there are three fatty acids attached, these are known as triglycerides. "Bread" and pastries from a "bread factory" often contain mono- and diglycerides as "dough conditioners." Can you figure out what these molecules would look like? The main distinction between fats and oils is whether they're solid or liquid at room temperature, and this, as we'll soon see, is based on differences in the structures of the fatty acids they contain.

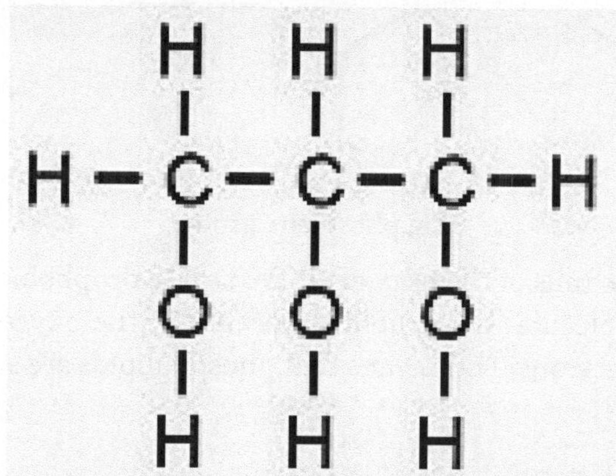

Lipids have different levels of saturation, a term that refers to the number of double bonds in the carbon chain. Fatty acids with no double bonds are saturated, those with one double bond are monounsaturated, and those with more than one double bond are polyunsaturated. Single bonds are stronger and less chemically reactive than double bonds, so the greater the number of double bonds, the greater the opportunity for the fatty acid to react with its chemical environment. It is this differential reactive capacity that makes the number of double bonds an important factor in human nutrition.

Saturated fatty acids are most prevalent in fats of animal origin, palm kernel oil, and coconut oil. Monounsaturated fats are highest in olive oil and canola oil but are also present in fats of animal origin. Polyunsaturated fats are highest in vegetable oils (with the exception of olive oil, which is more than 75 percent monounsaturated). In the context of a fat intake that does not exceed 35 percent of total calories, mono- and polyunsaturated fats should make up the majority of the fats consumed. Saturated fats are associated with higher cholesterol levels, so they should be minimized when possible. This is most easily achieved by reducing the consumption of animal fats, chocolate candies (often high in saturated tropical oils), fried foods, and high-fat dairy products.

7.2.2.1 Phospholipids

Phospholipids are made from glycerol, two fatty acids, and (in place of the third fatty acid) a phosphate group with some other molecule attached to its other end.

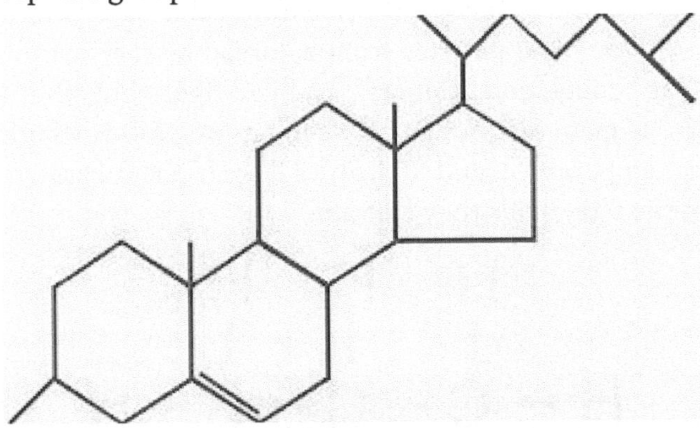

Fig phosphate group

The hydrocarbon tails of the fatty acids are still hydrophobic, but the phosphate group end of the molecule is hydrophilic because of the oxygens with all of their pairs of unshared electrons. This means that phospholipids are soluble in both water and oil.

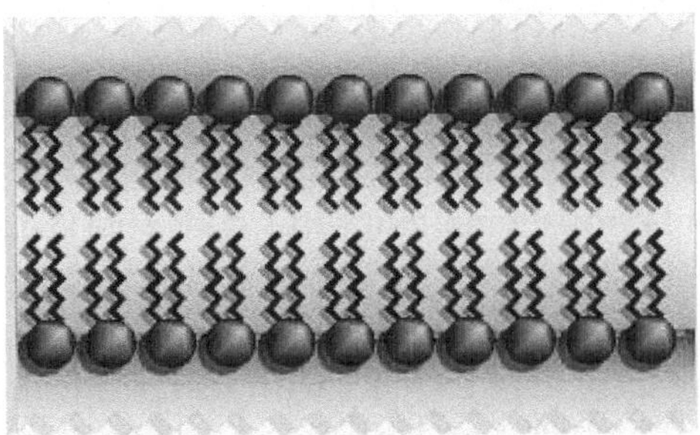

An emulsifying agent is a substance which is soluble in both oil and water, thus enabling the two to mix. A "famous" phospholipid is lecithin which is found in egg yolk and soybeans. Egg yolk is mostly water but has a lot of lipids, especially cholesterol, which are needed by the developing chick. Lecithin is used to emulsify the lipids and hold them in the water as an emulsion. Lecithin is the basis of the classic emulsion known as mayonnaise.

The cell membranes are made mostly of phospholipids arranged in a double layer with the tails from both layers "inside" (facing toward each other) and the heads facing "out" (toward the watery environment) on both surfaces.

7.2.3 Steroids

The general structure of cholesterol consists of two six-membered rings side-by-side and sharing one side in common, a third six-membered ring off the top corner of the right ring, and a five-membered ring attached to the right side of that. The central core of this molecule, consisting of four fused rings, is shared by all steroids, including estrogen (estradiol), progesterone, corticosteroids such as cortisol (cortisone), aldosterone, testosterone, and Vitamin D. In the various types of steroids, various other groups/molecules are attached around the edges.

Our bodies make about 2 g of cholesterol per day, and that makes up about 85% of blood cholesterol, while only about 15% comes from dietary sources. Cholesterol is the precursor to our sex hormones and Vitamin D. Vitamin D is formed by the action of UV light in sunlight on cholesterol molecules that have "risen" to near the surface of the skin. At least one source I read suggested that people not shower immediately after being in the sun, but wait at least ½ hour for the new Vitamin D to be absorbed deeper into the skin. Our cell membranes contain a lot of cholesterol (in between the phospholipids) to help keep them "fluid" even when our cells are exposed to cooler temperatures.

Lipoproteins are clusters of proteins and lipids all tangled up together. These act as a means of carrying lipids, including cholesterol, around in our blood. There are two main categories of lipoproteins distinguished by how compact/dense they are. LDL or low density lipoprotein is the "bad guy," being associated with deposition of "cholesterol" on the walls of someone's arteries. HDL or high density lipoprotein is the "good guy," being associated with carrying "cholesterol" out of the blood system, and is more dense/more compact than LDL.

Triglycerides

The majority of consumed lipids are triglycerides, which contain three fatty acids and a glycerol molecule. Fat is stored in the form of triglycerides, which we manufacture when excess energy is consumed. We store triglycerides in adipose tissues (groups of fat cells) and inside muscle cells (intramuscular tri glyceride), both of which are available as an energy source when needed. When fat is burned as a source of energy, the stored triglycerides are taken out of storage, and each molecule is cleaved into its component fatty acids and glycerol molecule. Each fatty acid can then be broken apart (two carbon units at a time) and thrown into the cellular furnaces for the creation of ATP to form heat and provide the energy for muscular work. This process is referred to as the beta-oxidative metabolic pathway because burning fat, besides requiring some carbohydrate for its complete oxidation, also requires oxygen.

Glycerol is a unique lipid that is burned like a carbohydrate rather than a fat and is also an effective humectant (it holds water). Some long-endurance athletes find that adding glycerol to water helps them retain more water (i.e., to superhydrate) than if they consumed water alone. In an extremely hot and humid environment, water loss is likely to be higher than the athlete's fluid-replacement capacity, so beginning a competition in a superhydrated state may provide some advantages. In studies of tennis and Olympic distance triathlon, athletes consuming glycerol in water prior to competition experienced some protective hyperhydration benefits when exercising in high heat. However, being superhydrated is likely to impart a degree of discomfort that requires some adaptation. Athletes who consume fluids containing glycerol often describe the feeling of holding extra body water as making them feel "like a water bag," or "heavy," or "stiff. " Still, they maintain that how they feel at the end of a race is more important than how they feel at the beginning, so adding glycerol to precompetition fluids has become a standard protocol for some athletes.

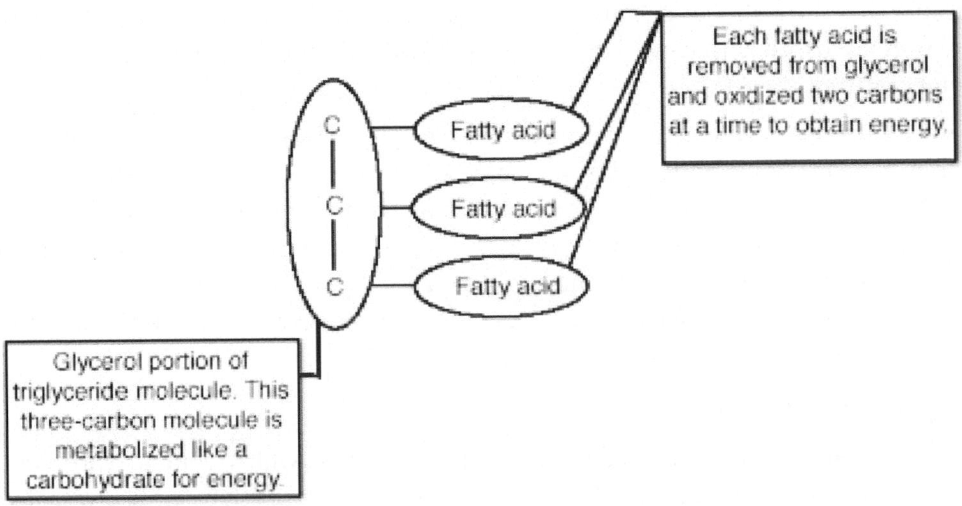

7.3 ESSENTIAL FATTY ACIDS

Linoleic (omega-6) and linolenic (omega-3) fatty acids are the essential fatty acids; although they are needed for metabolic processes, we are incapable of synthesizing them. The "omega-6" classification means that these fatty acids, which are polyunsaturated, have the last double bond six carbons from the end of the carbon chain. The "omega-3" classification means that these fatty acids have the last double bond three carbons from the end of the carbon chain. Linoleic acid is an essential part of lipid membranes and is required for normal skin health. Linolenic acid is necessary for neural function and growth. The AMDR for omega-6 andomega-3 fatty acids is 5 to 10 and .6 to 1.2 grams per day, respectively.

Both fatty acids are easily obtained from vegetable oils (corn, safflower, canola, and so on) and the oils of cold-water fish. Fish liver oils that are high in omega-3 acids have recently received much attention. These oils have been shown to reduce the ability of red blood cells to congregate, thereby decreasing the chance for an unwanted blood clot to form. This reduces the risk of a heart attack, which is most commonly caused by a clot formation in one of the major heart arteries. The oils from cold-water fish are the main source of the omega-3 fatty acids eicosapentanoic acid (EPA) and docosohexanoic acid (DHA). Even a once-weekly consumption of cold-water fish is sufficient to significantly reduce the risk of heart attack and stroke. Despite these findings, excessive intake of these fish oils may cause problems, including an increase in cellular oxidative damage. The best rule of thumb is to make fish consumption a regular part of the weekly diet so that supplemental intake of omega-3 fatty acids is unnecessary.

7.3.1 Lipids and Physical Activity

Even the leanest, healthiest athletes have a substantial energy pool of stored lipids. The average storage in adipose tissue ranges between 50,000 and 100,000 calories, or enough energy to walk or run 500 to 1,000 miles19 (800 to 1,600 kilometers) without a refueling stop. In addition, athletes store approximately 2,000 to 3,000 calories of lipids inside the muscle tissue. These lipids, which are stored in the form of triglycerides, are available as a fuel under the proper conditions of oxygen availability. Maximal fat oxidation occurs at 60 to 65 percent VO2 max, but at higher levels of VO2max there is insufficient oxygen to derive the majority of energy used from fat Catholicism.

Triglycerides stored in adipose tissue are cleaved into their component molecules of glycerol and fatty acids and transported to the blood plasma. The glycerol is available to all tissues for energy metabolism, and the free fatty acids are transported to working muscles where they are oxidized for energy. The triglycerides stored in working muscles are cleaved into glycerol and fatty acids, and the fatty acids can be oxidized for energy where they are resident. The glycerol can also be burned for energy in the working muscle or can be transported to the blood plasma as a source of energy for other tissues.

The lower the exercise intensity, the greater the proportion of fat burned to satisfy energy needs. As exercise intensity increases, the proportion of fat burned decreases and the proportion of carbohydrate burned increases. It is this basic reality that is behind why so many people do low-intensity activity to burn fat and lower body fat levels. However, the proportion of fat burned should not be confused by the total amount of fat burned at different intensities of physical activity. As exercise intensity increases, the total number of calories burned per unit of time also increases. Although there may be a decrease in the proportion of fat burned to satisfy total energy needs in higher-intensity activity, the total volume of fat burned is greater because the total energy requirement is higher. The take-away message from this metabolic reality is that athletes interested in lowering body fat should exercise at least as high as 65 percent of VO2max for the duration of their workouts to optimize the total mass of fat that is burned. Exercising at lower intensities burns a greater proportion of fat but less total fat than exercising at higher intensities.

7.4 BIOSYNTHESIS OF FATTY ACID

Carboxylic acids (R-COOH)	
Starter unit	**Extension unit**
Acetyl-CoA 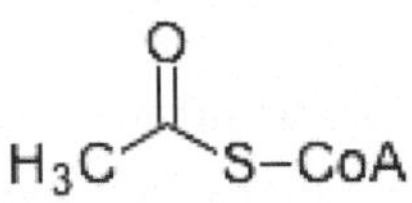 C00024	Malonyl-CoA 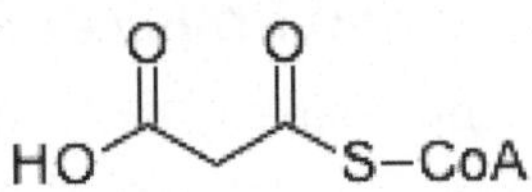 C00083

Initiation

Acyl transferase (AT)				Ketoacyl synthase (KS)
Acetyl-[acp] 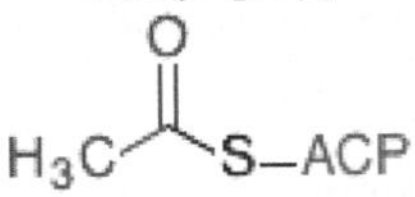	+	Malonyl-[acp] 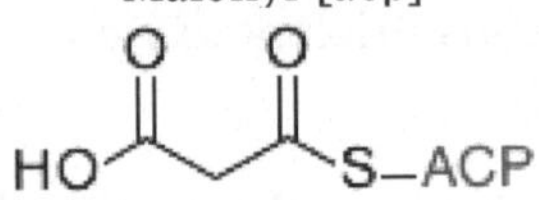	→	Acetoacetyl-[acp] 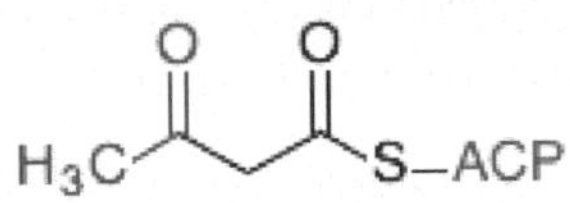

Elongation

Ketoacyl synthase (KS)	Keto reductase (KR)	Dehydratase (DH)	Enoyl reductase (ER)
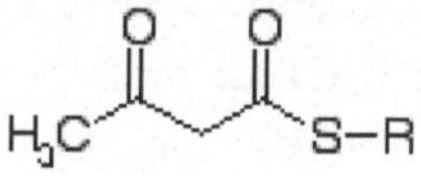 C05744	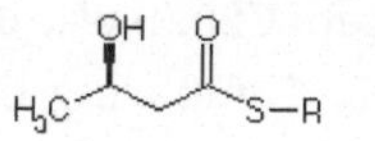 C04618	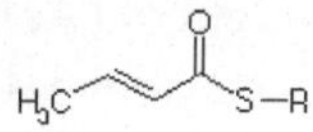 C04246	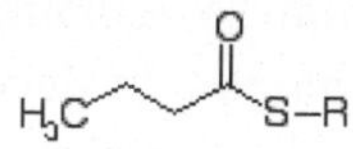 C05745
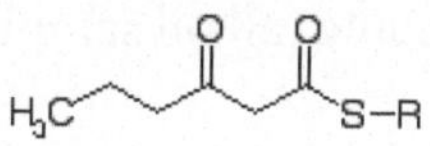 C05746	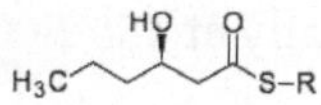 C05747	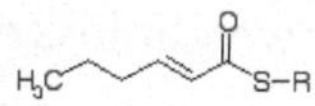 C05748	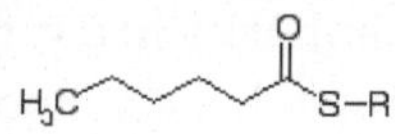 C05749
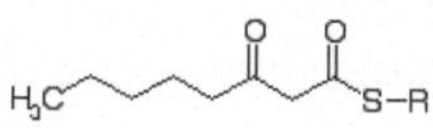 C05750	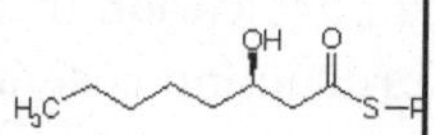 C04620	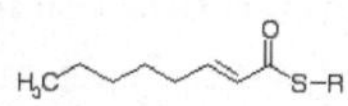 C05751	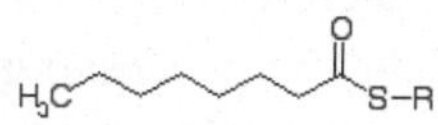 C05752
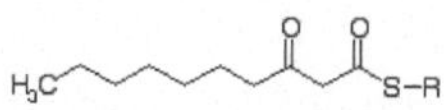 C05753	......		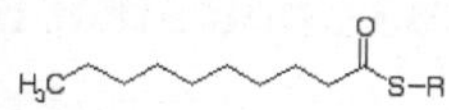 C05755
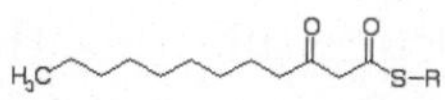 C05756	......		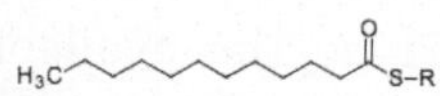 C05223
 C05759	......		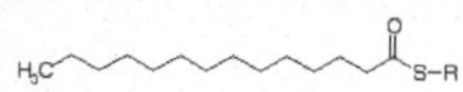 C05761
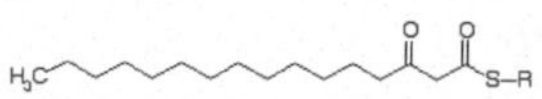 C05762	......		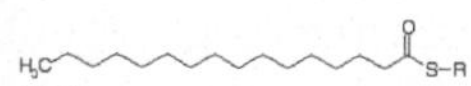 C05764

Termination

Thio reductase (TE)
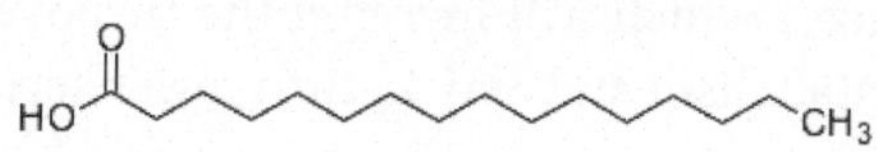 C00249 Palmitic acid (16:0)

7.5 DEGRADATION OF FATTY ACIDS

The oils and individual fatty acids, FAs (C14:0, C16:0, C18:0, C20:0, C16:1, C18:1, C18:2, C18:3, C20:2, C20:4, C22:6) were subjected to the Thermogravimetric Analyzer.

Samples (0.5-1 mg) was placed in the furnace and heated up to 600°C at the rate of 5°C/min. Weight loss (%) of oils and fatty acids were calculated based on the original weight of non-heated samples. Duplicate experiments were conducted.

Over 90% weight loss of all FAs (except C18:1 and C20:2) was observed at 250°C. Less than 2% remaining weight of FAs (except C20:2) was observed at 350°C. Saturated FAs completely decomposed above 400°C. Comparing C18:1 with C18:2 and C18:3, weight loss increased with increased double bonds. Between 200-550°C, weight loss of oils slightly increased with increased temperature, regardless of the purification process. There was, however, no distinct weight loss of all oils below 300°C. Oils with less impurity had greater weight loss especially at 450-500°C. At 550°C, all oil samples completely decomposed.

Knowledge of thermal stability or degradation (weight loss through evaporation or decomposition) of catfish oils during purification or cooking is essential for the design of processing units, analysis of production cost, and cooking quality if to be used as cooking or seasoning oil.

7.5.1 Cholestrol Biosynthesis and degradation

The formation of cholesterol by the liver, regulated by a type of homeostatic control, is influenced by the hepatic level offree cholesterol or glycogen, or both, and by the availability of the proper enzymes. Here the factors affecting the hepaticbiosynthesis of cholesterol are detailed, including the influence of dietary composition and fasting. The specific effects oncholesterol synthesis of feeding cholesterol, various steroids, saturated and unsaturated fats and thyroid hormone are reviewed. Also considered are the effects of radiation, glycolysis and various pharmacologic agents.

In the degradation of cholesterol, the chief metabolic end products are bile acids, although cholesterol may also be converted into sex hormones and adrenocortical hormones. The steps involved in the degradation of cholesterol to bileacids are reviewed, along with the various factors which increase or decrease cholesterol catabolism. Since degradation of cholesterol depends on the levels of bile salts in the liver, any factor which will increase the turnover of bile acids will also accelerate cholesterol catabolism and may in turn lower serum cholesterol.

Slightly less than half of the cholesterol in the body derives from biosynthesis de novo. Biosynthesis in the liver accounts for approximately 10%, and in the intestines approximately 15%, of the amount produced each day. Cholesterol synthesis occurs in the cytoplasm and microsomes from the two-carbon acetate group of acetyl-CoA.

The process has five major steps:

☆ Acetyl-CoAs are converted to 3-hydroxy-3-methylglutaryl-CoA (HMG-CoA)

- ☆ HMG-CoA is converted to mevalonate
- ☆ Mevalonate is converted to the isoprene based molecule, isopentenyl pyrophosphate (IPP), with the concomitant loss of CO2
- ☆ IPP is converted to squalene
- ☆ Squalene is converted to cholesterol.

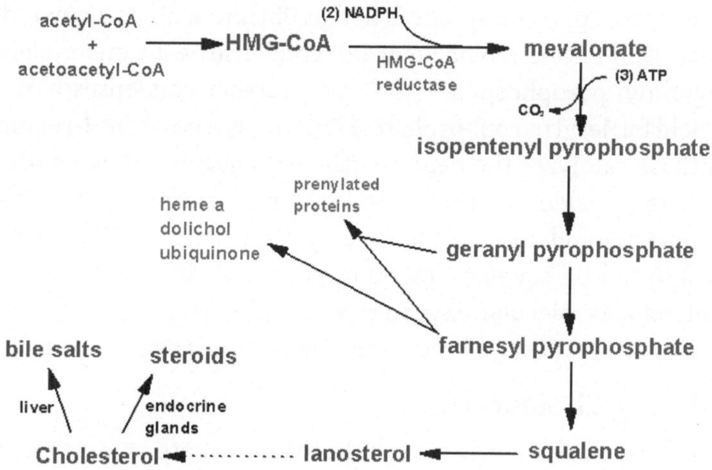

The acetyl-CoA utilized for cholesterol biosynthesis is derived from an oxidation reaction (eg, fatty acids or pyruvate) in the mitochondria and is transported to the cytoplasm by the same process as that described for fatty acid synthesis. Acetyl-CoA can also be derived from cytoplasmic oxidation of ethanol by acetyl-CoA synthetase. All the reduction reactions of cholesterol biosynthesis use NADPH as a cofactor. The isoprenoid intermediates of cholesterol biosynthesis can be diverted to other synthesis reactions, such as those for dolichol (used in the synthesis of N-linked glycoproteins, coenzyme Q (of the oxidative phosphorylation) pathway or the side chain of heme a. Additionally, these intermediates are used in the lipid modification of some proteins.

Acetyl-CoA units are converted to mevalonate by a series of reactions that begins with the formation of HMG-CoA. Unlike the HMG-CoA formed during ketone body synthesis in the mitochondria, this form is synthesized in the cytoplasm. However, the pathway and the necessary enzymes are the same as those in the mitochondria. Two moles of acetyl-CoA are condensed in a reversal of the thiolase reaction, forming acetoacetyl-CoA. Acetoacetyl-CoA and a third mole of acetyl-CoA are converted to HMG-CoA by the action of HMG-CoA synthase. HMG-CoA is converted to mevalonate by HMG-CoA reductase, HMGR (this enzyme is bound in the endoplasmic reticulum, ER). HMGR absolutely requires NADPH as a cofactor and two moles of NADPH are consumed during the conversion of HMG-CoA to mevalonate.

The reaction catalyzed by HMGR is the rate limiting step of cholesterol biosynthesis, and this enzyme is subject to complex regulatory controls. Mevalonate is then activated by three successive phosphorylations, yielding 5-pyrophosphomevalonate. In addition to activating mevalonate, the phosphorylations maintain its solubility, since otherwise it is insoluble in water. After phosphorylation, an ATP-dependent decarboxylation yields isopentenyl pyrophosphate, IPP, an activated isoprenoid molecule. Isopentenyl pyrophosphate is in equilibrium with its isomer, dimethylallyl pyrophosphate, DMPP. One molecule of IPP condenses with one molecule of DMPP to generate geranyl pyrophosphate, GPP. GPP further condenses with another IPP molecule to yield farnesyl pyrophosphate, FPP. Finally, the NADPH-requiring enzyme, squalene synthase catalyzes the head-to-tail condensation of two molecules of FPP, yielding squalene (squalene synthase also is tightly associated with the endoplasmic reticulum). Squalene undergoes a two step cyclization to yield lanosterol. The first reaction is catalyzed by squalene monooxygenase. This enzyme uses NADPH as a cofactor to introduce molecular oxygen as an epoxide at the 2,3 position of squalene. Through a series of 19 additional reactions, lanosterol is converted to cholesterol.

7.5.2 Regulating Cholesterol

Normal healthy adults synthesize cholesterol at a rate of approximately 1g/day and consume approximately 0.3g/day. A relatively constant level of cholesterol in the body (150 - 200 mg/dL) is maintained primarily by controlling the level of de novo synthesis. The level of cholesterol synthesis is regulated in part by the dietary intake of cholesterol. Cholesterol from both diet and synthesis is utilized in the formation of membranes and in the synthesis of the steroid hormones and bile acids. The greatest proportion of cholesterol is used in bile acid synthesis.

The cellular supply of cholesterol is maintained at a steady level by three distinct mechanisms:

- ☆ Regulation of HMGR activity and levels
- ☆ Regulation of excess intracellular free cholesterol through the activity of acyl-CoA:cholesterol acyltransferase, ACAT
- ☆ Regulation of plasma cholesterol levels via LDL receptor-mediated uptake and HDL-mediated reverse transport.

Regulation of HMGR activity is the primary means for controlling the level of cholesterol biosynthesis. The enzyme is controlled by four distinct mechanisms: feed-back inhibition, control of gene expression, rate of enzyme degradation and phosphorylation-dephosphorylation.

The first three control mechanisms are exerted by cholesterol itself. Cholesterol acts as a feed-back inhibitor of pre-existing HMGR as well as inducing rapid degradation of the enzyme. The latter is the result of cholesterol-induced polyubiquitination of HMGR and its degradation in the proteosome. This ability of cholesterol is a consequence of the sterol sensing domain, SSD of HMGR. In addition, when cholesterol is in excess the amount of mRNA for HMGR is reduced as a result of decreased expression of the gene. The mechanism by which cholesterol (and other sterols) affect the transcription of the HMGR gene is described below under regulation of sterol content.

Regulation of HMGR through covalent modification occurs as a result of phosphorylation and dephosphorylation. The enzyme is most active in its unmodified form. Phosphorylation of the enzyme decreases its activity. HMGR is phosphorylated by AMP-activated protein kinase, AMPK (this is not the same as cAMP-dependent protein kinase, PKA). AMPRK itself is activated via phosphorylation. The phosphorylation of AMPK is catalyzed by AMPK kinase.

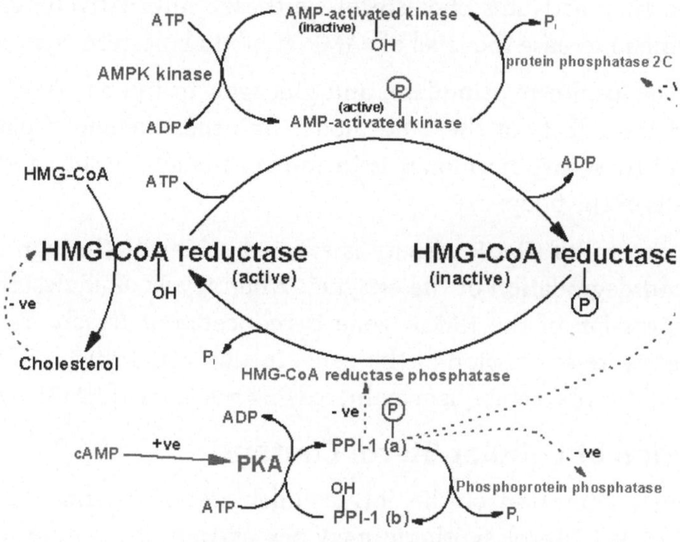

Fig phosphorylation of AMPK

Regulation of HMGR by covalent modification. HMGR is most active in the dephosphorylated state. Phosphorylation is catalyzed by AMP-activated protein kinase, AMPK, (used to be termed HMGR kinase), an enzyme whose activity is also regulated by phosphorylation. Phosphorylation of AMPK is catalyzed by AMPK kinase. Hormones such as glucagon and epinephrine negatively affect cholesterol biosynthesis by increasing the activity of the inhibitor of phosphoprotein phosphatase inhibitor-1, PPI-1. Conversely, insulin stimulates the removal of phosphates and, thereby, activates HMGR activity. Additional regulation of HMGR occurs through an inhibition of its' activity as well as of its' synthesis by elevation in intracellular cholesterol levels.

The activity of HMGR is additionally controlled by the cAMP signaling pathway. Increases in cAMP lead to activation of cAMP-dependent protein kinase, PKA. In the context of HMGR regulation, PKA phosphorylates phosphoprotein phosphatase inhibitor-1 (PPI-1) leading to an increase in its' activity. PPI-1 can inhibit the activity of numerous phosphatases including protein phosphatase 2C and HMG-CoA reductase phosphatase which remove phosphates from AMPRK and HMGR, respectively. This maintains AMPRK in the phosphorylated and active state, and HMGR in the phosphorylated and inactive state. As the stimulus leading to increased cAMP production is removed, the level of phosphorylations decreases and that of dephosphorylations increases. The net result is a return to a higher level of HMGR activity.

Since the intracellular level of cAMP is regulated by hormonal stimuli, regulation of cholesterol biosynthesis is hormonally controlled. Insulin leads to a decrease in cAMP, which in turn activates cholesterol synthesis. Alternatively, glucagon and epinephrine, which increase the level of cAMP, inhibit cholesterol synthesis.

The ability of insulin to stimulate, and glucagon to inhibit, HMGR activity is consistent with the effects of these hormones on other metabolic pathways. The basic function of these two hormones is to control the availability and delivery of energy to all cells of the body.

Long-term control of HMGR activity is exerted primarily through control over the synthesis and degradation of the enzyme. When levels of cholesterol are high, the level of expression of the HMGR gene is reduced. Conversely, reduced levels of cholesterol activate expression of the gene. Insulin also brings about long-term regulation of cholesterol metabolism by increasing the level of HMGR synthesis.

7.5.3 Regulation of Cellular Sterol Content

The continual alteration of the intracellular sterol content occurs through the regulation of key sterol synthetic enzymes as well as by altering the levels of cell-surface LDL receptors. As cells need more sterol they will induce their synthesis and uptake, conversely when the need declines synthesis and uptake are decreased. Regulation of these events is brought about primarily by sterol-regulated transcription of key rate limiting enzymes and by the regulated degradation of HMGR. Activation of transcriptional control occurs through the regulated cleavage of the membrane-bound transcription factor sterol regulated element binding protein, SREBP. As discussed above, degradation of HMGR is controlled by the ubiquitin-mediated pathway for proteolysis.

Sterol control of transcription affects more than 30 genes involved in the biosynthesis of cholesterol, triacylglycerols, phospholipids and fatty acids.

Transcriptional control requires the presence of an octamer sequence in the gene termed the sterol regulatory element, SRE-1. It has been shown that SREBP is the transcription factor that binds to SRE-1 elements. It turns out that there are 2 distinct SREBP gene, SREBP-1 and SREBP-2. In addition, the SREBP-1 gene encodes 2 proteins, SREBP-1a and SREBP-1c as a consequence of alternative exon usage. All 3 proteins are proteolytically regulated by sterols. Full-length SREBPs have several domains and are embedded in the membrane of the endoplasmic reticulum (ER). The N-terminal domain contains a transcription factor motif of the basic helix-loop-helix (bHLH) type that is exposed to the cytoplasmic side of the ER. There are 2 transmembrane spanning domains followed by a large C-terminal domain also exposed to the cytosolic side. When sterols are scarce, cleavage of the full-length SREBP takes place with the result being that the N-terminal bHLH motif is released into the cytosol. The bHLH domain then migrates to the nucleus to direct transcription. Conversely, when sterols are abundant, cleavage of SREBP is inhibited. To control the level of SREBP-mediated transcription, the soluble bHLH domain is itself subject to rapid proteolysis.

The cleavage of SREBP is carried out by 2 distinct enzymes, one of which is regulated by sterols. The regulated cleavage occurs in the lumenal loop between the 2 transmembrane domains. This cleavage is catalyzed by site-1 protease, S1P. High sterol content blocks the activity of S1P. The second cleavage, catalyzed by site-2 protease, S2P, occurs in the first transmembrane span, leading to release of active SREBP. In order for S2P to act on SREBP, site-1 must already have been cleaved.

Additional studies on sterol-regulated gene expression demonstrated that cleavage of SREBP by S1P is controlled by the level and action of an additional protein termed, SREBP cleavage-activating protein, SCAP. SCAP is a large protein also found in the ER membrane and contains at least 8 transmembrane spans. The C-terminal portion, which extends into the cytosol, has been shown to interact with the C-terminal domain of SREBP. This C-terminal region of SCAP contains 4 motifs called WD40 repeats. The WD40 repeats are required for interaction of SCAP with SREBP. Interestingly, the N-terminus of SCAP, including membrane spans 2-6, resembles HMGR which itself is subject to sterol-stimulated degradation. This shared motif is called the sterol sensing domain, SSD. Several proteins whose functions involve sterols also contain the SSD. These include patched, an important development regulating receptor whose ligand, hedgehog, is modified by attachment of cholesterol and the Neimann Pick C1 (NPC1) protein which is involved in cholesterol transport in the secretory pathway. NPC1 is one of several genes whose activities, when disrupted, lead to severe neurological dysfunction.

The function of SCAP is to positively stimulate S1P-mediated cleavage of SREBP. The function of sterols is to inhibit this positive action of SCAP. The activity of SCAP involves movement from the ER to the Golgi and back. Because the C-terminus of SCAP interacts with SREBP, movement of SCAP takes SREBP along for the ride. When sterols are low, SCAP and SREBP move to the Golgi. This transit is required for SREBP cleavage as S1P is Golgi-localized. When sterols are high, movement of SCAP is halted. Thus, the overall effect of sterols is to regulate the ability of SCAP to present SREBP to S1P.

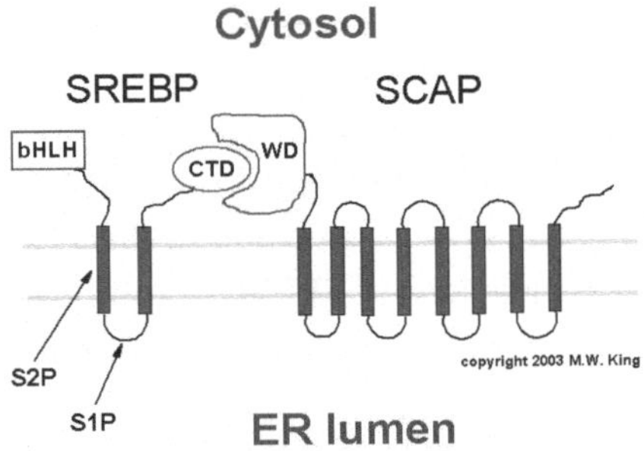

Fig effect of sterols

7.5.4 Structure of conjugated cholic acids

Within the liver the carboxyl group of primary and secondary bile acids is conjugated via an amide bond to either glycine or taurine before their being resecreted into the bile canaliculi. These conjugation reactions yield glycoconjugates and tauroconjugates, respectively. The bile canaliculi join with the bile ductules, which then form the bile ducts. Bile acids are carried from the liver through these ducts to the gallbladder, where they are stored for future use. The ultimate fate of bile acids is secretion into the intestine, where they aid in the emulsification of dietary lipids. In the gut the glycine and taurine residues are removed and the bile acids are either excreted (only a small percentage) or reabsorbed by the gut and returned to the liver. This process of secretion from the liver to the gallbladder, to the intestines and finally reabsorbtion is termed the enterohepatic circulation.

7.6 DISORDERS OF LIPID METABOLISM

Fats (lipids) are an important source of energy for the body. The body's store of fat is constantly broken down and reassembled to balance the body's energy needs with the food available. Groups of specific enzymes help the body break down and process fats. Certain abnormalities in these enzymes can lead to the buildup of specific fatty substances that normally would have been broken down by the enzymes. Over time, accumulations of these substances can be harmful to many organs of the body. Disorders caused by the accumulation of lipids are called lipidoses. Other enzyme abnormalities prevent the body from converting fats into energy normally. These abnormalities are called fatty acid oxidation disorders.

7.6.1 GAUCHER'S DISEASE

Gaucher's disease is caused by a buildup of glucocerebrosides in tissues. Children who have the infantile form usually die within a year, but children and adults who develop the disease later in life may survive for many years.

In Gaucher's disease, glucocerebrosides, which are a product of fat metabolism, accumulate in tissues. Gaucher's disease is the most common lipidosis. The disease is most common among Ashkenazi (Eastern European) Jews. Gaucher's disease leads to an enlarged liver and spleen and a brownish pigmentation of the skin. Accumulations of glucocerebrosides in the eyes cause yellow spots called pingueculae to appear. Accumulations in the bone marrow can cause pain and destroy bone.

Type 1, the chronic form of Gaucher's disease, is the most common. It results in an enlarged liver and spleen and bone abnormalities. Most commonly diagnosed during adulthood, type 1 Gaucher's disease may lead to severe liver disease, including increased risk of bleeding from the stomach and esophagus and liver cancer. Nervous system problems can also occur.

Type 2, the infantile form, usually causes death in the first year of life. Affected infants have an enlarged spleen and severe nervous system problems.

Type 3, the juvenile form, can begin at any time during childhood. Children with type 3 disease have an enlarged liver and spleen, bone abnormalities, and slowly progressive nervous system problems. Children who survive to adolescence may live for many years.

Many people with Gaucher's disease can be treated with enzyme replacement therapy, in which enzymes are given by vein, usually every 2 weeks. Enzyme replacement therapy is most effective for people who do not have nervous system complications.

7.6.2 TAY-SACHS DISEASE

- ☆ Tay-Sachs disease is caused by a buildup of gangliosides in the tissues. This disease results in early death.
- ☆ In Tay-Sachs disease, gangliosides, which are products of fat metabolism, accumulate in tissues. The disease is most common among families of Eastern European Jewish origin.
- ☆ Before conception, parents can find out whether they carry the gene that causes the disease. During pregnancy, Tay-Sachs disease can be identified in the fetus by chorionic villus sampling or amniocentesis.

7.6.3 NIEMANN-PICK DISEASE

- Niemann-Pick disease is caused by a buildup of sphingomyelin or cholesterol in the tissues. This disease causes many neurologic problems.
- In Niemann-Pick disease, the deficiency of a specific enzyme results in the accumulation of sphingomyelin (a product of fat metabolism) or cholesterol. Niemann-Pick disease has several forms, depending on the severity of the enzyme deficiency, which determines how much sphingomyelin or cholesterol accumulates.
- In the most severe form (type A), children fail to grow normally and have several neurologic problems. These children usually die by age 3. Children with type B disease develop fatty growths in the skin, areas of dark pigmentation, and an enlarged liver, spleen, and lymph nodes.
- Some forms of Niemann-Pick disease can be diagnosed in the fetus by chorionic villus sampling or amniocentesis. After birth, the diagnosis can be made by a liver biopsy (removal of a tissue specimen for examination under a microscope). None of the types of Niemann-Pick disease can be cured, and children tend to die of infection or progressive dysfunction of the central nervous system.

7.6.4 FABRY'S DISEASE

- Fabry's disease is caused by a buildup of glycolipid in tissues. This disease causes skin growths, pain in the extremities, poor vision, recurrent episodes of fever, and kidney or heart failure.
- In Fabry's disease, glycolipid, which is a product of fat metabolism, accumulates in tissues. Because the defective gene for this rare disorder is carried on the X chromosome, the full-blown disease occurs only in males.
- The accumulation of glycolipid causes noncancerous (benign) skin growths (angiokeratomas) to form on the lower part of the trunk. The corneas become cloudy, resulting in poor vision. A burning pain may develop in the arms and legs, and children may have episodes of fever.
- Children with Fabry's disease eventually develop kidney failure and heart disease, although most often, they live into adulthood. Kidney failure may lead to high blood pressure, which may result in stroke.
- Fabry's disease can be diagnosed in the fetus by chorionic villus sampling or amniocentesis.

☆ The disease cannot be cured or even treated directly, but researchers are investigating a treatment in which the deficient enzyme is replaced by transfusion.

☆ Treatment consists of taking analgesics to help relieve pain and fever or anticonvulsants. People with kidney failure may need a kidney transplant.

7.7 FATTY ACID OXIDATION DISORDERS

Fatty acid oxidation disorders are caused by a lack or deficiency of the enzymes needed to break down fats, resulting in delayed mental and physical development.

Several enzymes help break down fats so that they may be turned into energy. An inherited defect or deficiency of one of these enzymes leaves the body short of energy and allows breakdown products, such as acyl-CoA, to accumulate. The enzyme most commonly deficient is medium chain acyl-CoA dehydrogenase (MCAD). Other enzyme deficiencies include short chain acyl-CoA-dehydrogenase deficiency (SCAD), long chain-3-hydroxyacyl-CoA-deficiency (LCHAD), and trifunctional protein deficiency (TFP).

7.7.1 MCAD Deficiency

This disorder is one of the most common inherited disorders of metabolism, particularly among people of Northern European descent. Symptoms usually develop between birth and age 3. Children are most likely to develop symptoms if they go without food for a period of time (which depletes other sources of energy) or have an increased need for calories because of exercise or illness. The level of sugar in the blood drops significantly, causing confusion or coma. Children become weak and may have vomiting or seizures. Over the long term, children have delayed mental and physical development, an enlarged liver, heart muscle weakness, and an irregular heartbeat.

Since 2007, nearly every state in the United States has screened all newborns for MCAD with a blood test. Immediate treatment is with glucose given by vein. For long-term treatment, children must eat often, never skip meals, and consume a diet high in carbohydrates and low in fats. Supplements of the amino acid carnitine may be helpful. The long-term outcome is generally good.

GLOSSARY

Acid-Base Balance: In medicine, the state of having the right amount of acid and base in the blood and other body fluids. Keeping a normal acid-base balance is important for the body to work the way it should. Also called acid-base equilibrium.

Amino Acid: A large organic molecule that is the basic building block of proteins. There are 20 different amino acids that link together in various order to form proteins. The order of amino acids is determined by the genetic sequence.

Antioxidant: A substance that protects cells from the damage caused by free radicals (unstable molecules made by the process of oxidation during normal metabolism). Free radicals may play a part in cancer, heart disease, stroke, and other diseases of aging. Antioxidants include beta-carotene, lycopene, vitamins A, C, and E, and other natural and manufactured substances.

Calorie: A unit commonly used to measure energy content of foods and beverages as well as energy use (expenditure) by the body.

Carbohydrate: One of three macronutrients in food that provide calories, or "energy" for the body. There are several types of carbohydrate: sugars, sugar alcohols, starches, and ietary fiber.

Cardiovascular Disease: Heart disease as well as diseases of the blood vessel system (arteries, capillaries, veins) that can lead to heart attack, chest pain (angina), or stroke.

Cholesterol: A natural sterol present in all animal tissues. Free cholesterol is a component of cell membranes and serves as a precursor for steroid hormones (estrogen, testosterone, aldosterone), and for bile acids.

Enzyme: A protein that speeds up chemical reactions in the body

Fat: Fatty acids that have one double bond and are usually liquid at room temperature.

Fat: Unsaturated fatty acids that are structurally different from the unsaturated fatty acids that occur naturally in plant foods.

Grain: Grains and grain products with the bran and germ removed; any grain product that is not a whole-grain product.

Hormones: Chemicals produced by glands in the body and circulated in the bloodstream.

Minerals: Inorganic substances that are required by the body in relatively small amounts (also called micronutrients) for normal growth and activity.

Nutrient: A substance in food that contributes to growth and health; nutrients provide energy, cell building and structural materials, and agents that regulate body chemistry

Oils: Fats that are liquid at room temperature. Oils come from many different plants and some fish.

Protein: One of three macronutrients in food that provide calories, or "energy," for the body.

Sugars: Composed of one unit (a monosaccharide, such as glucose or fructose) or two joined units (a disaccharide, such as lactose or sucrose).

Starch: Many glucose units linked together into long chains. Examples of foods containing starch include beans and peas (e.g., garbanzo beans, kidney beans, lentils, and split peas), grains (e.g., barley, brown rice, corn, oats, and wheat), and vegetables (e.g., carrots and potatoes)

Vitamins: Organic substances that are required by the body in relatively small amounts (also called micronutrients) for normal growth and activity.

CHAPTER-8

NUCLEIC ACIDS

8.1 NUCLEIC ACID AS GENETIC MATERIAL

Nucleic acids are the basic unit of inheritance, they constitute all the genetic material of living things. This genetic information determines how individual cells and entire organisms develop and function (by encoding for protein synthesis).

8.1.1 Structure of Nucleotides

The monomeric unit of a nucleic acid is a nucleotide, which is comprised of a 5C-sugar, a phosphate group and a nitrogenous base. The carbon atoms of the sugar molecule are numbered (1 - 5), the base connects to the 1'-C while the phosphate connects to the 5'-C of the sugar.

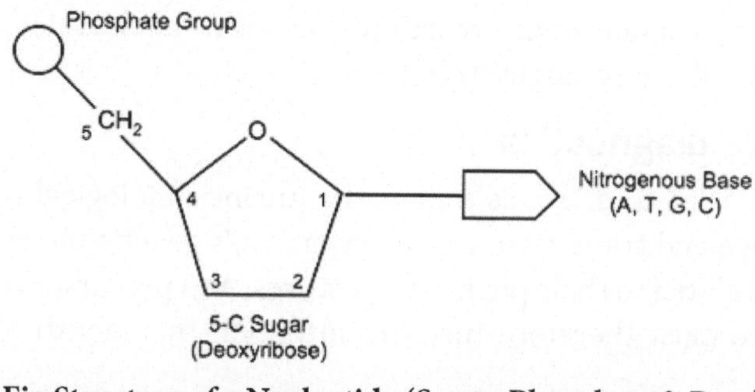

Fig Structure of a Nucleotide (Sugar, Phosphate & Base)

The nucleotide will contain one of four different nitrogenous bases, and when nucleotides are joined into chains, the order of these bases will determine the specific genetic information encoded by that sequence of nucleic acid. The four bases in DNA are:

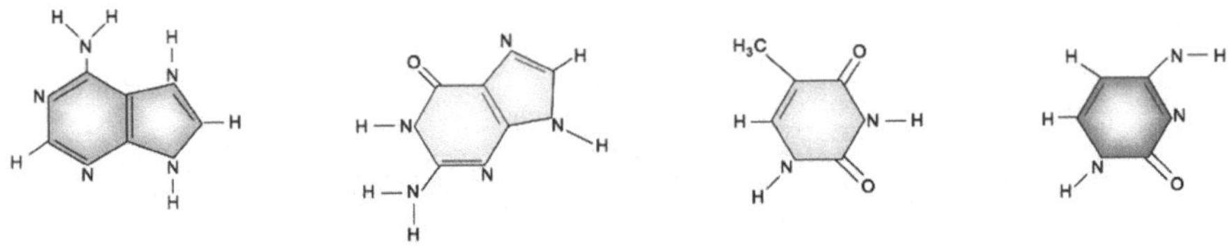

Fig Adenine, Guanine, Thymine and Cytosine

- ☆ Adenine (A) and guanine (G) are purines (double ring bases)
- ☆ Thymine (T) and cytosine (C) are pyrimidines (single ring bases)

8.1.2 Biosynthesis of Nucleic Acids

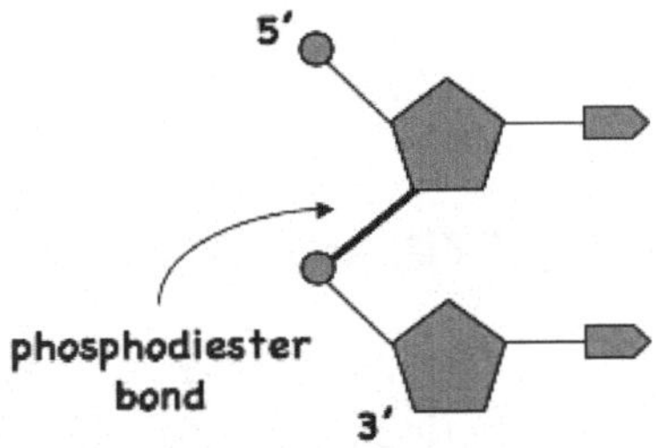

Fig Nucleotide

- ☆ Nucleotides a linked into a single strand via a condensation reaction
- ☆ The phosphate group (attached to the 5'-C of the sugar) joins with the hydroxyl (OH) group attached to the 3'-C of the sugar
- ☆ This results in a phosphodiester bond between the two nucleotides and the formation of a water molecule
- ☆ Successive condensation reactions between nucleotides results in the formation of a long single strand

8.1.3 Molecular diagnostics

The methods used to do so, as well as the intrinsic biological roles of DNA and RNA in the storage and transmission of an organism's genetic information, turn out to be intimately related to their physical structures. As a precursor to understanding MDx methods we must therefore turn our attention this month to DNA and RNA structure.

Table. Names of DNA bases, nucleosides, and nucleotides

Base	Nucleoside	Nucleotide
Adenine	deoxyadenosine	deoxyadenylate
Guanine	deoxyguanosine	deoxyguanylate
Cytosine	doxycytidine	deoxycytidylate
Thymine	deoxythymidine	deoxythymidylate

DNA is best considered in terms of its constituent subparts. The reader is probably aware that there are four such subparts, and that each of the four consists of a heterocyclic nitrogen-containing molecule (the nitrogenous base, or just "base," in context); the base covalently linked to a 5 carbon sugar (2′ – deoxyribose), with the base plus sugar being referred to as a "nucleoside"; and the nucleoside with the covalent addition of one, two, or three phosphate groups to the 5′ carbon, which respectively make up a nucleoside mono-, di-, or triphosphate (collectively referred to as a "nucleotide"). Table gives the names for each of these 12 molecules, and Fig shows the structures for all four bases, plus an example of deoxyadenylate with the base, nucleoside, and nucleotide portions labelled.

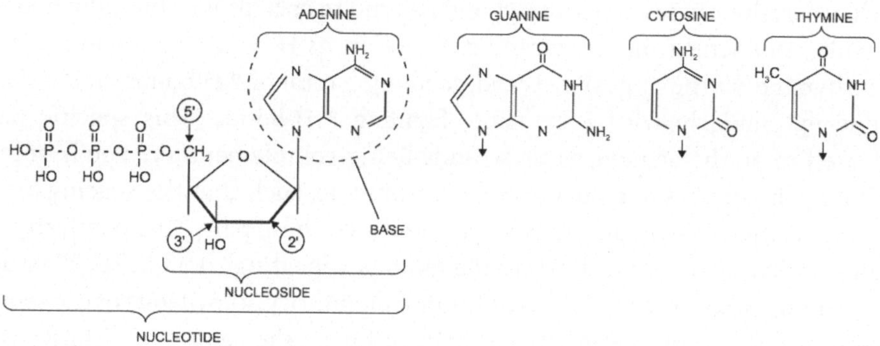

Fig structures for four bases

For the other nucleosides and nucleotides, the reader can just envision exchanging the base part of the structure. In the above fig, on deoxyribose portion, carbons labelled 2′, 3′, and 5′ have been labelled. The 2′ position here doesn't have a sugar (thus it's 2′-deoxyribose), the reactive parts of the molecule are the 3′ -OH, and the phosphate groups are attached to the 5′ position. Then look at the bases, and notice the –NH, N, and =O groups either in the ring or exocyclic (sticking out). These form a set of H-bond donors and acceptors, as discussed last month. It's important to note that each base has its own unique pattern of these. Note also that two of the bases (adenine and guanine) are large, two-ring structures (purines), while cytosine and thymine are smaller, single-ring entities (pyrimidines).

A DNA molecule is initially composed of a string of nucleoside monophosphates (NMPs) held together by covalent bonds between the 3′ –OH of one subunit, and the innermost (alpha) phosphate group on the 5′ end of another NMP (the other two phosphate groups are released in the joining reaction). The 3′ end of this NMP is then attached to the 5′ alpha phosphate of another NMP, and so on, to create a long, single stranded chain of NMPs. Together the sugar and phosphate parts form a repeating backbone of this strand, with the bases projecting off to the side. The strand still has exactly one free and reactive 5′ end and at the opposite end, a free, reactive 3′ –OH group. This is known as the polarity of the strand, and by convention we consider single DNA strands from their 5′ towards their 3′ ends. As the backbone is exactly the same repeating unit, the NMPs can be strung together in any order, and we generally refer to this order by just naming the 5′ to 3′ order of the base components such as 5′-ACTGAATGCA-3′ or 5′-TGGATC-3′. Chains of roughly 6 to 50 or so joined nucleotides are generically called oligonucleotides, while longer chains are polynucleotides.

It is in such polynucleotides that DNA gains its biological utility. If there are two strands of DNA and they orient their polarities in opposite directions (antiparallel) to each other, their bases can interact and H-bond to each other—but only under one strict rule. An adenine on one strand can line up its H-bond donors and acceptors with a thymine on the opposite strand, forming a total of 2 H bonds; and a guanine can line up similarly with a cytosine, forming 3 H-bonds. This specific pairing is referred to as the sequences or strands being complementary. You'll note that each base pair consists of a purine and a pyrimidine, such that the spacing between backbones is the same on an A:T base pair and a G:C base pair. This regularity along two long antiparallel strands leads to the famous, repeating "double helix" structure of DNA familiar to everyone. Because of complementarity, knowledge of the sequence of bases on one strand automatically defines for us the sequence of bases on the antiparallel second strand. For instance, from our examples in the paragraph above, the antiparallel strands would have to be 5′-TGCATTCAGT-3′ and 5′-GATCCA-3′.

When two DNA single strands can align with each other, each of the H-bonds that can form between the correct A:T and G:C base pairs has a small attractive energy. Individually, each H-bond formed is quite weak, but when a number of them occur in series such as between two complementary strands, the effect is additive and the result is a non-covalent but strong net force holding the strands together. If the strands are not complementary, then the H-bond donors and acceptors don't line up, and this collective attractive force isn't present to hold them together. Additionally, if two non-complementary strands try to pair up any two purines, they're physically too large to both fit between the backbones; and if two pyrimidines are placed across from each other, they're too small and their respective H-bond donors and acceptors not only don't match up, but are also too far apart to effectively interact. Only a purine—pyrimidine pair has the right size, and only A:T and G:C sets match each other's H-bonding capacities.

Asking what holds two strands of DNA together presupposes there's some other force trying to pull the strands apart. There is, and that force is none other than heat. Heat, after all, is just random motion of molecules, and the force of this vibrational motion depends on temperature. At physiological temperature ranges, the thermal energy in any one single nucleotide is more than the energy it can find in its complementary H-bonds. Thus, a free adenylate in a cell doesn't automatically pair up with the first thymidylate it passes; their vibrational energies carry them apart as they surpass their net attractive forces. As two complementary strands get longer, the additive nature of the H-bond energies comes into play, and at some total number of potential H-bonds between two single DNA strands, their energy surpasses the temperature dependent vibrational energies, and the two strands stick together. What this exact temperature is depends on the length of the DNA molecules (longer chains of nucleotides have more bases and thus more potential H-bonds) and also the sequence; as mentioned above an A:T base pair has two H-bonds while a G:C base pair has three, so sequences with more "G:C content" stick together more strongly than sequences with more A:T base pairs.

The process whereby complementary DNA strands find each other and pair up their H-bonds to effectively overcome thermal energies trying to shake them apart is referred to as annealing, and two DNA strands H-bonded to each other are referred to as annealed. The attractive force between two annealed strands is a fixed value based on total number of H-bonds; the thermal energy acting to try to vibrate the strands apart of course is not fixed, and can be manipulated by heating or cooling of the environment. If the environment is heated to the point where thermal energy overcomes the attractive force between two DNA strands, vibration wins out and the strands separate or denature. The temperature at which thermal energy exactly matches H-bond energy for a given DNA paired sequence, and the strands hover between annealing and denaturation, is known as the melting temperature or Tm of the sequence. To get a sense of the energy and temperature ranges here, a DNA sequence of about equal numbers of A:T and G:C base pairs, about 13 base pairs in total length, has a Tm of around 37°C, while a similar random sequence of about 20 bases has a Tm of about 51°C. Making sequences longer or more G:C rich raises the Tm, but there is a limit; any DNA sequence will effectively denature as its environment approaches 100°C and the aqueous environment starts a liquid-to-gas phase change. In fact, since the solvent isn't pure water, the limit temperature at which essentially all DNA sequences denature is slightly lower, in the range of 95°C. We'll see later how MDx techniques utilize the processes of annealing and denaturation by temperature manipulation as the core of multiple methods.

8.2 DNA- DOUBLE HELIX STRUCTURE

8.2.1 Introduction

A DNA double helix consists of two spiral chains of deoxyribonucleic acid. The shape is similar to that of spiral staircase. DNA is a nucleic acid composed of nitrogenous bases (adenine, cytosine, guanine and thymine), a five-carbon sugar (deoxyribose), and phosphate molecules. The nucleotide bases of DNA represent the stair steps of the staircase and the deoxyribose and phosphate molecules form the sides of the staircase.

DNA is coiled into chromosomes and tightly packed in the nucleus of our cells. The twisting aspect of DNA is a result of interactions between the molecules that comprise DNA and water. The nitrogenous bases that compose the steps of the twisted staircase are held together by hydrogen bonds. Adenine is bonded with thymine (A-T) and guanine pairs with cytosine (G-C). These nitrogenous bases are hydrophobic, meaning that they lack an affinity for water. Since the cell cytoplasm and cytosol contain water-based liquids, the nitrogenous bases want to avoid contact with cell fluids. The sugar and phosphate molecules that form the sugar-phosphate backbone of the molecule are hydrophilic. This means they are water loving and have an affinity for water.

DNA is arranged such that the phosphate and sugar backbone are on the outside and in contact with fluid, while the nitrogenous bases are in the inner portion of the molecule.

In order to further prevent the nitrogenous bases from coming into contact with cell fluid, the molecule twists to reduce space between the nitrogenous bases and the phosphate and sugar strands. The fact that the two DNA strands that form the double helix are anti-parallel helps to twist the molecule as well.

Anti-parallel means that the DNA strands run in opposite directions ensuring that the strands fit tightly together. This reduces the potential for fluid to seep between the bases.

8.2.2 DNA Replication and Protein Synthesis

The double helix shape allows for DNA replication and protein synthesis to occur. In these processes, the twisted DNA unwinds and opens to allow a copy of the DNA to be made. In DNA replication, the double helix unwinds and each separated strand is used to synthesize a new strand. As the new strands form, bases are paired together until two double helix DNA molecules are formed from a single double helix DNA molecule. DNA replication is required for the processes of mitosis and meiosis to occur.

In protein synthesis, the DNA molecule is transcribed to produce an RNA version of the DNA code known as messenger RNA (mRNA). The messenger RNA molecule is then translated to produce proteins. In order for DNA transcription to take place, the DNA double helix must unwind and allow an enzyme called RNA polymerase to transcribe the DNA. RNA is also a nucleic acid, but contains the base uracil instead of thymine. In transcription, guanine pairs with cytosine and adenine pairs with uracil to form the RNA transcript. After transcription, the DNA closes and twists back to its original state.

8.2.3 Features

☆ The double helix of DNA has these features:

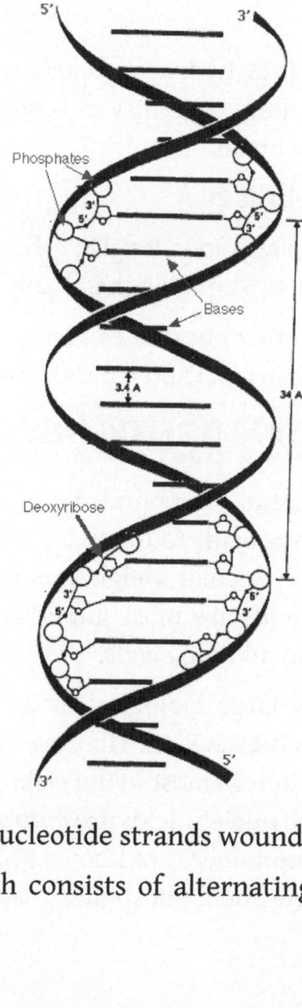

☆ It contains two polynucleotide strands wound around each other.

☆ The backbone of each consists of alternating deoxyribose and phosphate groups.

- ☆ The phosphate group bonded to the 5' carbon atom of one deoxyribose is covalently bonded to the 3' carbon of the next.
- ☆ The two strands are "antiparallel"; that is, one strand runs 5′ to 3′ while the other runs 3′ to 5′.
- ☆ The DNA strands are assembled in the 5′ to 3′ direction [More] and, by convention, we "read" them the same way.
- ☆ The purine or pyrimidine attached to each deoxyribose projects in toward the axis of the helix.
- ☆ Each base forms hydrogen bonds with the one directly opposite it, forming base pairs (also called nucleotide pairs).
- ☆ 3.4 Å separate the planes in which adjacent base pairs are located.
- ☆ The double helix makes a complete turn in just over 10 nucleotide pairs, so each turn takes a little more (35.7 Å to be exact) than the 34 Å shown in the diagram.
- ☆ There is an average of 25 hydrogen bonds within each complete turn of the double helix providing a stability of binding about as strong as what a covalent bond would provide.
- ☆ The diameter of the helix is 20 Å.
- ☆ The helix can be virtually any length; when fully stretched, some DNA molecules are as much as 5 cm (2 inches!) long.
- ☆ The path taken by the two backbones forms a major (wider) groove (from "34 A" to the top of the arrow) and a minor (narrower) groove (one below).

8.3 PROPERTIES AND FUNCTION OF NUCLEIC ACIDS

Nucleic acids are large organic compounds found in the chromosomes of living cells and viruses. They are strong acids found in the nucleus of the cells. The nucleic acid polymers are with high molecular weights as high as 100,000,000 grams per mole. With proteins, nucleic acids are most important biological macromolecules. They are found in abundance in all living cells.

Nucleic acids are essential large biological molecules for all forms of life. The nucleic acids include the DNA and the RNA. They are the hereditary determinants of living organisms. They are present in most living cells either in free state or bound to proteins as nucleoproteins. The nucleic acids are biopolymers with mononucleotides ad their repeating units. The monomers are known as nucleotides, they are made up of three units: a sugar, an amine and a phosphate group.

8.3.1 Properties of Nucleic Acids

Properties of nucleic acids are as follows:

- ☆ Nucleic acids (DNA and RNA) are long polymers made of repeating units of nucleotides.
- ☆ Nucleotide units are made of phosphate-sugar-nitrogenous base units.
- ☆ The nitrogenous bases found in DNA are adenine, guanine, cytosine and thymine.
- ☆ Adenine and guanine are purine bases while cytosine and thymine are pyrimidnes.
- ☆ In RNA, the thymine bases id replaced by uracil which is also a pyrimidine.
- ☆ The nucleotides are linked with phosphodiester bonds.
- ☆ They are linked by a phosphate group on the 5th position of sugar residue becomes linked to 3' hydroxyl group of the preceeding sugar molecule.
- ☆ The double stranded model of the DNA was worked out by Watson and Crick in 1953.
- ☆ The double helix model consists of two strands wound around a central axis with the bases stacked inside.
- ☆ The order of the strand is in opposite directions, i.e., from 5' to 3' direction in one and 3'to 5' direction in the other.
- ☆ The bases stacked in the center of the helix as they interact with each other through weak hydrogen bonds.
- ☆ Hydrogen bonds are weaker than covalent bonds, they are continually forming and disassociating.
- ☆ In the double stranded nucleic acids, the adenine form hydrogen bonds only with thymine (or uracil) molecule. While cytidine will only form hydrogen bonds with guanine.
- ☆ Hence, in a a given strand of DNA, the amount of adenine is always equal to the amount of thymine, and the amount of cytidine always equals the amount of guanine, in a given species.
- ☆ The per cent of the G-T and the A-T is variable from species to species.
- ☆ The base pairs form a flat plain in the helix, the adenine forming two hydrogen bonds with thymine, and the cytidine forming three bonds with guanine.

- ☆ Using the concept of base pairing, all the enzymes and substrates necessary, the two DNA strands when copied separately, wherever there is adenine in the original strand, the duplicated strand will have thymidine, and guanine would be matched to cytosine.
- ☆ After replication, each original strand (parent strand), is paired with a new (daughter) strand. This type of replication is known as semi-conservative mode of duplication.
- ☆ The double stranded DNA can be denatured by alkaline conditions or heat.
- ☆ Small stretches of polynucleotides will anneal to larger single stranded DNA molecule if the DNA sequence matches, by base pairing.

8.3.2 Functions

Functions of nucleic acids are:

- ☆ The main functions is store and transfer genetic information.
- ☆ To use the genetic information to direct the synthesis of new protein.
- ☆ The deoxyribonucleic acid is the storage for place for genetic information in the cell.
- ☆ DNA controls the synthesis of RNA in the cell.
- ☆ The genetic information is transmitted from DNA to the protein synthesizers in the cell.
- ☆ RNA also directs the production of new protein by transmitting genetic information to the protein building structures.
- ☆ The function of the nitrogenous base sequences in the DNA backbone determines the proteins being synthesized.
- ☆ The function of the double helix of the DNA is that no disorders occur in the genetic information if it is lost or damaged.
- ☆ RNA directs synthesis of proteins.
- ☆ m-RNA takes genetic message from RNA.
- ☆ t-RNA transfers activated amino acid, to the site of protein synthesis.
- ☆ r-RNA are mostly present in the ribosomes, and responsible for stability of m-RNA.

8.4 RNA- STRUCTURE AND FUNCTIONS OF M-RNA, T-RNA AND R-RNA

8.4.1 Introduction

RNA is typically single stranded and is made of ribonucleotides that are linked by phosphodiester bonds. A ribonucleotide in the RNA chain contains ribose (the pentose sugar), one of the four nitrogenous bases (A, U, G, and C), and a phosphate group. The subtle structural difference between the sugars gives DNA added stability, making DNA more suitable for storage of genetic information, whereas the relative instability of RNA makes it more suitable for its more short-term functions. The RNA-specific pyrimidine uracil forms a complementary base pair with adenine and is used instead of the thymine used in DNA.

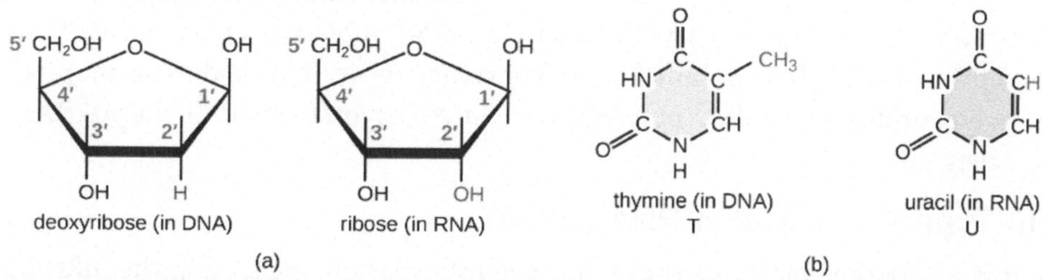

Fig. (a) Ribonucleotides with pentose sugar ribose. (b) RNA with pyrimidine uracil

Cells access the information stored in DNA by creating RNA to direct the synthesis of proteins through the process of translation. Proteins within a cell have many functions, including building cellular structures and serving as enzyme catalysts for cellular chemical reactions that give cells their specific characteristics. The three main types of RNA directly involved in protein synthesis are messenger RNA (mRNA), ribosomal RNA (rRNA), and transfer RNA (tRNA).

8.4.2 Types

There are three types of ribonucleic acid (RNA) such as

- ☆ Messenger RNA (mRNA)
- ☆ Transfer RNA (tRNA)
- ☆ Ribosomal RNA (rRNA).

RNA Type # 1. Messenger RNA (mRNA):

It carries the genetic message from the chromosomes (DNA) to the ribosome (site of protein synthesis). This RNA is formed in the nucleus (or nuclear zone in prokaryotes) containing complementary base sequence to a part of one strand of DNA (gene).

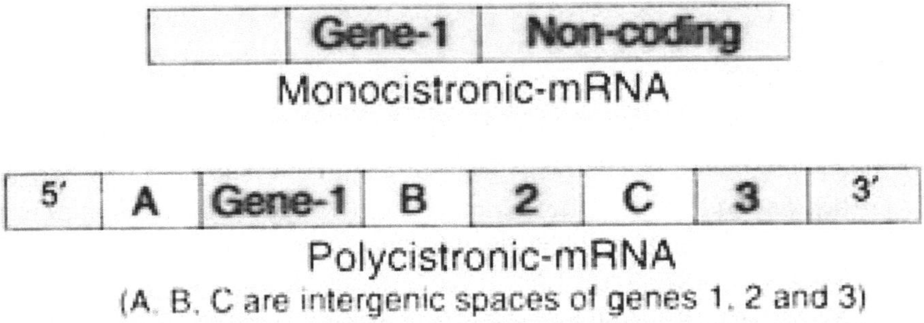

Fig Monocistronic-mRNA and Polycistronic-mRNA

The base sequence in the mRNA specifies the amino acid sequence in the polypeptide chains. In prokaryotes a single mRNA molecule codes either for one polypeptide chain, hence called monocistronic, or it may code for more than one polypeptide, thus called polycistronic. In eukaryotes, most of the mRNAs are monocistronic.

RNA Type # 2. Transfer RNA (tRNA):

It is an adapter molecule that reads the information encoded in the mRNA and accordingly transfers the appropriate amino acid to the growing polypeptide chain during protein synthesis. There is at least one kind of tRNA for each amino acid. Some amino acids have two or more specific tRNAs. At least 32 tRNA's are required to recognize all the amino acid codons (64 — 3 = 61).

Most cells have to about 40-50 distinct tRNAs. Some tRNA's can recognize more than one codon (due to wobbling). tRNAs have between 73 and 93 nucleotides in their structures. Eight or more nucleotides are unusual modified bases which are methylated derivatives of the principal bases. Most tRNA's have guanylate (pG) at 5′ end and have the base sequence CCA at 3′ end.

These are single stranded and there is maximum intra- chain base pairing which gives it a clover leaf shape structure, with four arms. The four arms of tRNA are:

- ☆ Amino acyl arm (AA arm)
- ☆ Anticodon arm
- ☆ Dihydrouridine arm (DHU arm) and
- ☆ Ribothymidine – pseudouridine arm (TΨC Arm).

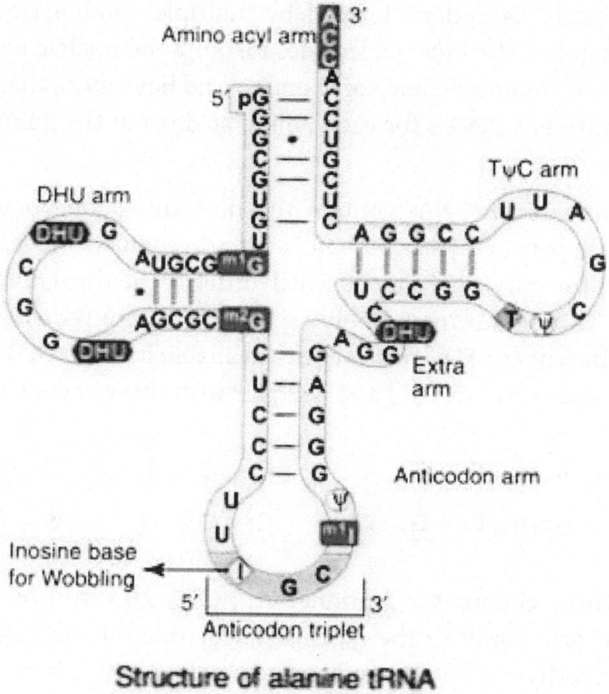

Fig Structure of Alanine tRNA

The amino acyl arm carries a specific amino acid esterified to its carboxyl group at the 2′ or 3′ hydroxyl group of the adenosine present at the 3′ end of the tRNA. The anticodon arm contains the anticodon, i.e., a sequence of three nucleotide bases complementary to the genetic code in mRNA for that particular amino acid. The two RNAs are paired anti-parallel, i.e., the first base of the codon (5′ → 3′) pairs with the third base of the anticodon (3′ → 5′) creating codon-anticodon interaction as shown.

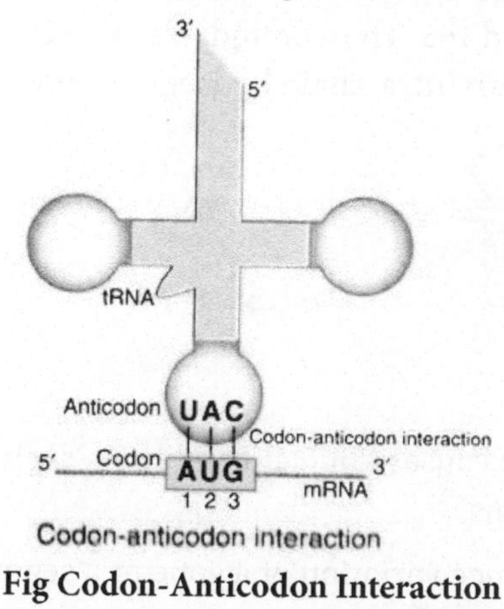

Fig Codon-Anticodon Interaction

There are totally 64 codons formed by multiple combinations of any three nucleotide bases out of the four nucleotides forming the nucleic acids. At least one codon specifies one amino acid and some amino acid have more than one codon, but the number of different tRNA's for each amino acid is not the .fame as the number of its codons.

This is because some tRNAs contain the first anticodon nucleotide inosinate (I), which can base pair by hydrogen bonds with three nucleotides, viz., U, C and A. Therefore, all those amino acid codons which differ at the third nucleotide will have the same tRNA. This is known as wobbling, e.g., there are six codons for arginine, but only four different types of tRNAArg are available in the cell. One of these tRNA contains the anticodon (3′) GCI (5′) and will pair with three codons for arginine.

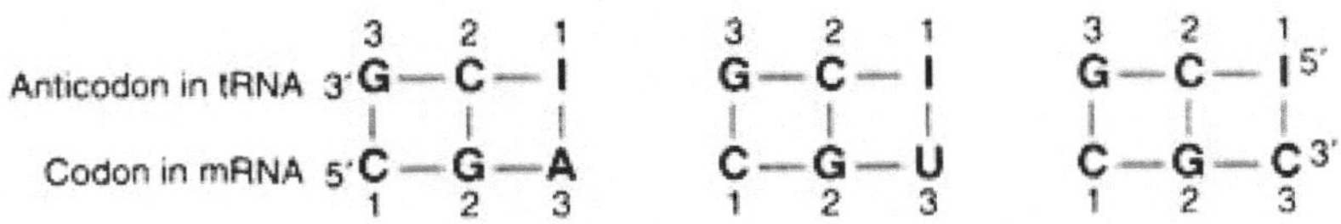

The other three codons for arginine, viz., CGG, AGA and AGG have different tRNAs. The DHU arm contains the unusual nucleotide dihydrouracil and the TΨC arm contains ribothymidine (T) and pseudouridine (Ψ) which has an unusual carbon-carbon bond between the base and pentose sugar. The functions of DHU arm is recognition of its proper aminoacyl-tRNA synthetase and TΨC arm is involved in binding of the aminoacyl-tRNA to the ribosomal surface.

RNA Type # 3. Ribosomal RNA (rRNA):

Ribosomal RNA molecules in association with proteins forms the seat of protein synthesis or the protein synthesis machinery called the ribosome. The RNAs are designated depending upon their sedimentation coefficients. The rRNAs found in prokaryotes are 5S, 23S and 16S. Those found in eukaryotes are 5S, 18S 28S, and 5.8S. rRNA is single stranded with intra-chain hydrogen bonding.

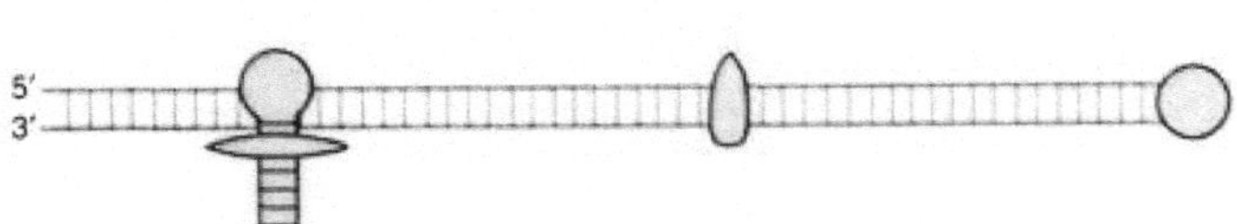

Ribosomal RNA

GLOSSARY

Adenovirus: A virus that causes clinical conditions such as the common cold and respiratory infections.

Allele: A particular sequence variation of a gene or a segment of a chromosome.

Alternative splicing: The processing of an RNA transcript into different mRNA molecules by including some exons and excluding others.

Amino acid: The basic building block of a protein. There are 20 different amino acids commonly found in proteins. The genetic code specifies the sequence of amino acids in a protein.

Amplification: The repeated copying of a DNA sequence.

Annealing: The hydrogen bonding between complementary DNA (or RNA) strands to form a double helix.

Annotation: The process of locating genes, their coding regions, and their functions.

Anticodon: A 3-base sequence in a tRNA molecule that base-pairs with its complementary codon in an mRNA molecule.

Assembly: Putting sequenced fragments of DNA into their correct order along the chromosome.

Autosome: Any of the numbered chromosomes that are not involved with sex determination.

Bacterial Artificial Chromosome (BAC): A chromosome-like structure constructed using recombinant-DNA technology. It is used to clone large DNA inserts (100 to 300 kb) into E. coli cells.

Base pair: Two nitrogenous bases (adenine and thymine/uracil or guanine and cytosine) held together by weak hydrogen bonds.

Base-pair substitution: A type of mutation where one base pair is replaced with a different one; also called a point mutation.

Base: One of five chemicals (also called nitrogenous bases) found in nucleic acids (DNA and RNA): adenine, thymine, guanine, cytosine, and uracil.

Bioethics: The study of ethical issues raised by the developments in the life science technologies.

BLAST (Basic Local Alignment Search Tool): A computer program that searches for sequence similarities. It can be used to identify homologous genes in different organisms.

Candidate gene: A gene that is suspected of being associated with a particular disease.

Carrier: A person who is heterozygous for a mutation associated with a genetic disease. Usually, a carrier does not display symptoms of the disease but may pass the mutation on to offspring.

cDNA (complementary DNA): A DNA molecule synthesized from an mRNA molecule. They can be used experimentally to determine the sequence of an mRNA.

Centromere: The compact region near the center of a chromosome.

Chromosome: A rod-like structure found in the cell nucleus. It consists of one long DNA molecule with its associated proteins.

Clone: 1) a genetically identical copy of an individual cell or organism;

Cloning vector: A DNA molecule, such as a modified plasmid or virus, that can be used to clone other DNA molecules in a suitable host cell. Cloning vectors must be able to replicate in the host cell and must possess restriction enzyme cut sites that allow the DNA molecules targeted for cloning to be inserted and retrieved.

Coding DNA or region: A sequence of DNA that is translated into protein; also called exons (in eukaryotes).

Codon: A three-base sequence in a DNA or mRNA molecule that specifies a specific amino acid or termination signal; the basic unit of the genetic code.

Combined DNA Index System (CODIS): A database maintained by the FBI. It includes DNA profiles of convicted offenders in the United States.

Comparative genomics: The process of learning about human genetics by comparing human DNA sequences with those from other organisms.

Consanguineous: Marriage or mating among related individuals.

Conserved sequence: A DNA (or amino acid) sequence that has remained relatively unchanged throughout evolution. Such a sequence is under selective pressure and therefore resistant to change.

Contig: contiguous sequence of DNA created by assembling shorter, overlapping sequenced fragments of a chromosome (whether natural or artificial, as in BACs). A list or diagram showing an ordered arrangement of cloned overlapping fragments that collectively contain the sequence of an originally continuous DNA.

Cosmid: A cloning vector derived from a bacterial virus. It can accommodate about 40 kb of inserted DNA.

Cycle sequencing: A DNA sequencing technique that combines the chain termination method developed by Fred Sanger with aspects of the polymerase chain reaction.

Deletion: A type of mutation caused by the loss of one or more adjacent base pairs from a gene.

Deoxyribose: The five-carbon sugar component of DNA. It has one less hydroxyl group than ribose, the sugar component of RNA.

Dideoxynucleotides (ddNTPs): Synthetic nucleotides lacking both 2' and 3' hydroxyl groups. They act as chain terminators during DNA sequencing reactions.

Directed sequencing: Successively sequencing DNA from adjacent stretches of chromosome.

DNA (deoxyribonucleic acid): The hereditary material that exists as a double-stranded helical molecule made up of a deoxyribose sugarphosphate backbone and the four nitrogenous bases, named adenine (A), cytosine (C), guanine (G), and thymine (T).

DNA chip: A microarray of oligonucleotides or cDNA clones fixed on a surface. They are commonly used to test for sequence variation in a known gene, or to profile gene expression in an mRNA preparation.

DNA ligase: An enzyme able to form a phosphodiester bond between adjacent but unlinked nucleotides in a double helix.

DNA polymerase: An enzyme that adds bases to a replicating DNA strand.

DNA probe: A chemically synthesized, often radioactively labeled, segment of DNA used to visualize a genomic sequence of interest by hydrogen-bonding to its complementary sequence.

DNA replication: The process of replicating a double-stranded DNA molecule.

Dominant: In human genetics, it describes any trait that is expressed in the heterozygous condition.

Draft sequence: DNA sequence with lower accuracy than finished sequence. Some segments may be missing or in the wrong orientation.

Duplication: A chromosome rearrangement that duplicates a given region of DNA. Duplications may occur in tandem or the sequences may be inverted.

Electropherogram: Sequence data produced by an automated DNA sequencing machine. The data are a series of colored peaks where each peak represents one of the four different DNA bases.

Euchromatin: The gene-rich areas of a chromosome.

Eukaryote: An organism whose cells contain a nucleus in which the chromosomes are located, as compared to prokaryotes, in which the chromosome is found throughout the cytoplasm.

Exon: A segment of a gene that codes for a portion of a protein. Exons are interspersed with noncoding introns.

Expressed Sequence Tag (EST): A short DNA sequence from a coding region that is used to identify a gene.

Finished sequence: Sequence produced to an accuracy of no more than 1 error in 10,000 bases. Finished sequences are in the proper orientation and have little or no gaps.

Frame shift mutation: A type of mutation characterized by insertions or deletions that change the identities of the codons following the mutation. Often this creates stop codons that cause premature termination of the protein.

Functional genomics: The study of genomes to determine the biological function of all the genes and their products.

Gene expression: The process by which a gene is transcribed into RNA and then translated into a protein.

Gene: A region of DNA that can encode one or more polypeptides or RNA products.

Genetic code: The mapping between the set of 64 possible three-base codons and the amino acids or stop codons specified by each of the triplets.

Genome: The complete DNA content of an organism.

Genomics: The comprehensive study of whole sets of genes and their interactions rather than single genes or proteins.

Genotype: The allelic composition of an individual for one or more genes.

Germ cell: A haploid egg or sperm cell.

Haploid: A single set of chromosomes found in the sperm and eggs of an animal or the pollen and egg of a plant.

Haplotype: A specific combination of alleles or sequence variations that are likely to be inherited together.

Heritability: The proportion of variation in a trait among individuals in a population that can be attributed to genetic effects.

Heterochromatin: The compact, gene-poor regions of a chromosome. They contain many repeated sequences.

Heterozygote: An individual that possesses two different alleles for a given gene.

Homologous genes: Genes having similar structures and functions.

Homozygote: An individual that possesses two identical alleles for a given gene.

Insertion: A type of mutation caused by the addition of one or more adjacent base pairs to a gene.

Intron: A gene region that is not translated into protein. Introns are interspersed with coding regions called exons.

Karyotype: A photomicrograph that arranges a cells chromosomes to show their number, size, and type.

Kilobase (kb): A unit of DNA length corresponding to one thousand bases.

Linkage: The proximity of two or more loci (especially genes) on a chromosome.

Locus: The position on a chromosome where a gene, or some other sequence, is located.

Megabase: A unit of DNA length corresponding to 1 million bases.

Messenger RNA (mRNA): An RNA molecule that serves as a template for the synthesis of a protein.

Microsatellite: Repetitive stretches of short DNA sequences that are used as markers to track the inheritance of genes.

Missense mutation: A type of mutation that results in the substitution of one type of amino acid for another in a given location in a polypeptide chain.

Multiple sequence alignment: A bioinformatics tool that compares multiple DNA or amino acid sequences and aligns them to highlight their similarities.

Mutagen: A chemical or physical agent that interacts with DNA to promote the appearance of mutations.

Mutation: A change in a DNA sequence with respect to a reference sequence.

Nonsense mutation: A type of mutation that changes an amino acid codon to one of the three stop codons, resulting in a shorter and usually nonfunctional protein.

Nucleic acid: A large polymer consisting of a linear stretch of nucleotides, as in DNA and RNA.

Nucleotide: The building block of a nucleic acid, consisting of a five-carbon sugar covalently bonded to a nitrogenous base (adenine, thymine, guanine, cytosine, or uracil) and a phosphate group.

Oligonucleotide: A short, synthetically made stretch of single-stranded DNA.

Pedigree: A family tree describing the occurrence of heritable traits across as many generations as possible.

Phenotype: The physical traits of an organism that are determined by the genotype.

Phylogenetic tree: A tree-like diagram that depicts the evolutionary relationships between different organisms.

Plasmid: A small circular DNA molecule found in bacteria that replicates independently of the chromosome. Plasmids are used as cloning vectors.

Point mutation: A type of mutation that involves changing a single base in a DNA sequence.

Polygenic inheritance: Describing a type of inheritance where more than one gene contributes to a phenotype.

Polymerase chain reaction (PCR): An enzyme-mediated technique that allows specific DNA sequences to be amplified.

Polymorphism: A relatively common DNA sequence variation within a population at a given chromosomal location.

Predisposition: The condition of having a genotype that increases the risk for developing a genetic disease, if other environmental conditions are present.

Prokaryote: A cell or organism without a membrane-bound nucleus. Bacteria are prokaryotes.

Protein: A macromolecule consisting of one or more amino acid chains. Proteins carry out most of the cell functions.

Proteome: The full complement of proteins produced by a genome.

Proteomics: The study of the full set of proteins encoded by a genome and their interactions.

Pseudogene: A DNA sequence similar to that of an active gene. Pseudogenes have collected mutations that render them inactive.

Reading frame: The way an mRNA is read as a series of triplet codons during translation. There are three possible reading frames for any mRNA, and the correct reading frame is set by recognition of the AUG initiation codon.

Recessive: A trait is recessive if it is manifest only in the homozygous condition.

Recombinant DNA: A DNA molecule consisting of DNA from different sources; made using restriction enzymes and DNA ligase.

Recombination (also called crossing over): The process by which two homologous chromosomes exchange genetic material during the formation of eggs and sperm.

Repetitive DNA: A DNA sequence that is present in many identical or similar copies in the genome. The copies can be tandemly repeated or dispersed.

Restriction enzyme: An endonuclease isolated from bacteria that recognizes and cuts a DNA sequence at a specific sequence. They are used in genetic engineering.

Restriction fragment length polymorphism (RFLP): Differences in DNA sequence on homologous chromosomes that result in restriction fragments of varying lengths that can be detected using DNA probes.

Retrovirus: A virus that carries its genetic material as RNA, rather than DNA. Retroviruses use reverse transcriptase to insert their genetic material into the chromosomes of infected cells.

Reverse transcriptase: An RNA-dependent DNA polymerase isolated from retrovirus infected cells. It synthesizes a complementary DNA from an RNA template.

Ribonucleic acid (RNA): A type of nucleic acid consisting of nucleotides with a ribose sugar and the nitrogenous bases adenine, cytosine, guanine, and uracil (A, C, G, and U); usually single-stranded; functions in protein synthesis and as the genome of some viruses. Common types include mRNA, tRNA, and rRNA.

RNA splicing: The process by which introns are removed and exons are spliced together from an RNA transcript to produce an mRNA molecule.

Sense strand: The DNA strand of a gene that is complementary in sequence to the template (antisense) strand, and identical to the transcribed mRNA sequence (except that DNA contains T where RNA has U). Gene sequences found in databases are always of the sense strand, in the 5' to 3' direction.

Sequence tagged site (STS): A short stretch of DNA whose sequence occurs once in the genome and whose location is known. It serves as a landmark used in the mapping and assembly of a genome.

Short tandem repeat (STR): A short (2 to 5 bases) DNA sequence that repeats itself in tandem. STRs are used in DNA profiling.

Silent (synonymous) mutation: A type of mutation that changes a codon but does not alter the amino acid encoded. Such mutations may still have effects on mRNA splicing or stability.

Single Nucleotide Polymorphism (SNP): A common single-base pair variation in a DNA sequence.

Somatic cell: Any cell in a multicellular organism except a sperm or egg cell.

Splice site: The point in the sequence of the RNA transcript at which splicing takes place. Splice sites are found at exon-intron boundaries.

Start (or initiation) codon: The first AUG (methionine) codon to be used by the ribosome at the start of translation.

Stop codon: The codons UAA, UGA, or UAG, which cause the termination of translation.

Transcription factor: A protein that binds to regulatory regions and controls gene expression.

Transcriptome: The full complement of activated genes as represented by the set of mRNAs and transcripts, in a particular tissue at a particular time.

Transformation: The process of introducing foreign DNA into a cell, or of a cell becoming cancerous.

Translocation: A type of chromosome aberration in which a sequence of DNA from one chromosome is moved to another chromosome.

Transposon: A short DNA sequence that has the ability to move from one chromosomal position to another.

Vector: A DNA molecule that replicates independently in a host cell. It is used to ferry a foreign DNA sequence into a cell to be cloned.

Yeast Artificial Chromosome (YAC): A yeast-derived DNA sequence that can be spliced with a large fragment of foreign DNA and inserted into yeast cells to bo be amplified and sequenced.

CHAPTER- 9

CONCEPT OF GENE, GENOME AND GENE EXPRESSION

9.1 GENE

A gene is a union of genomic sequences encoding a coherent set of potentially overlapping functional products. In particular, the word gene was first used by Wilhelm Johannsen in 1909, based on the concept developed by Gregor Mendel in 1866 (Mendel 1866). The word was a derivative of pangene, which was used by Hugo De Vries for entities involved in pangenesis, Darwin's hypothetical mechanism of heredity. Johannsen called a gene the "special conditions, foundations and determiners which are present [in the gametes] in unique, separate and thereby independent ways by which many characteristics of the organism are specified". The etymology of the term derives from the Greek genesis ("birth") or genos ("origin").

9.1.1 Gene as physical molecule

The fact that heredity has a physical, molecular basis was demonstrated by the observation that X rays could cause mutations. Griffith's demonstration that something in virulent but dead Pneumococcus strains could be taken up by live nonvirulent Pneumococcus and transform them into virulent bacteria was further evidence in this direction. It was later shown that this substance could be destroyed by the enzyme DNase. In 1955, Hershey and Chase established that the substance actually transmitted by bacteriophage to their progeny is DNA and not protein.

Moreover, the idea that a gene's product is a diffusible substance underlies the complementation test that was used to define genes in the early years of bacteriology. A practical view of the gene was that of the cistron, a region of DNA defined by mutations that in trans could not genetically complement each other.

9.1.2 Gene as transcribed code

It was the solution of the three-dimensional structure of DNA by Watson and Crick in 1953 that explained how DNA could function as the molecule of heredity. Base pairing explained how genetic information could be copied, and the existence of two strands explained how occasional errors in replication could lead to a mutation in one of the daughter copies of the DNA molecule.

From the 1960s on, molecular biology developed at a rapid pace. The RNA transcript of the protein-coding sequences was translated using the genetic code into an amino acid sequence. Francis Crick summarized the flow of information in gene expression as from nucleic acid to protein (the beginnings of the "Central Dogma"). However, there were some immediate exceptions to this: It was known that some genes code not for protein but for functional RNA molecules such as rRNA and tRNA. In addition, in RNA viruses the gene is made of RNA. The molecular view of the gene that developed through the 1960s can be summarized in general terms to be a code residing on nucleic acid that gives rise to a functional product.

9.1.3 Gene as open reading frame (ORF) sequence pattern

The development of cloning and sequencing techniques in the 1970s, combined with knowledge of the genetic code, revolutionized the field of molecular biology by providing a wealth of information on how genes are organized and expressed. The first gene to be sequenced was from the bacteriophage MS2, which was also the first organism to be fully sequenced. The parallel development of computational tools led to algorithms for the identification of genes based on their sequence characteristics. In many cases, a DNA sequence could be used to infer structure and function for the gene and its products. This situation created a new concept of the "nominal gene," which is defined by its predicted sequence rather than as a genetic locus responsible for a phenotype. The identification of most genes in sequenced genomes is based either on their similarity to other known genes, or the statistically significant signature of a protein-coding sequence. In many cases, the gene effectively became identified as an annotated ORF in the genome.

9.2 GENOME

A genome is an organism's complete set of DNA, including all of its genes. Each genome contains all of the information needed to build and maintain that organism.

In humans, a copy of the entire genome—more than 3 billion DNA base pairs—is contained in all cells that have a nucleus.

A genome is the complete set of genetic information in an organism. It provides all of the information the organism requires to function. In living organisms, the genome is stored in long molecules of DNA called chromosomes. Small sections of DNA, called genes, code for the RNA and protein molecules required by the organism. In eukaryotes, each cell's genome is contained within a membrane-bound structure called the nucleus. Prokaryotes, which contain no inner membranes, store their genome in a region of the cytoplasm called the nucleoid. The full range of RNA molecules expressed by a genome is known as its transcriptome, and the full assortment of proteins produced by the genome is called its proteome.

Each cell in the body, for example, a skin cell or a liver cell, contains this same set of instructions:

- ☆ The instructions in our genome are made up of DNA.
- ☆ Within DNA is a unique chemical code that guides our growth, development and health.
- ☆ This code is determined by the order of the four nucleotide bases that make up DNA, adenine, cytosine, guanine and thymine, A, C, G and T for short.
- ☆ DNA has a twisted structure in the shape of a double helix.
- ☆ Single strands of DNA are coiled up into structures called chromosomes?.
- ☆ Your chromosomes are located in the nucleus? within each cell.
- ☆ Within our chromosomes, sections of DNA are "read" together to form genes?.
- ☆ Genes control different characteristics such as eye colour and height.
- ☆ All living things have a unique genome?.
- ☆ The human genome is made of 3.2 billion bases of DNA but other organisms have different genome sizes.

There are 23 pairs of chromosomes in the human genome. Between 1990 and 2003, all twenty-three pairs were fully sequenced through an international research undertaking known as the Human Genome Project. The study and analysis of genomes is called genomics.

9.2.1 Whole Genome Sequencing

Whole genome sequencing is the mapping out of a person's unique DNA. Your genome is the unique blueprint for your body. Sometimes, because of new or inherited genetic mutations, your genes can cause a disease or increase your risk for disease. By sequencing your genome, health professionals can look at the unique variations found in your genes. Some of it matters. Some doesn't matter. Some is still unknown or uncertain.

It is most often used in medical research and is beginning to be used more in clinical practice. For example, a doctor or genetic counselor could use whole genome sequencing to see if a patient has a genetic disorder or is at risk for a disease. Whole genome sequencing results can be placed into 3 categories: Single-gene disorders, multi-factorial disorders, and the pharmacogenomic profile.

- ☆ Single-gene disorders (sometimes also called Mendelian disorders) are diseases that are caused by a mutation in the DNA for one gene. An example of these diseases is Sickle Cell Anemia .
- ☆ Multi-factorial disorders are diseases associated with DNA changes in more than one gene. This often includes diseases like obesity and diabetes and often is highly influenced by your environment.
- ☆ Pharmacogenomic profiles use an individual's genetic code to determine how they will respond to a drug so that a doctor can prescribe the correct amount. This is an example of personalized medicine.

Whole genome sequencing is not your average diagnostic test. A blood pressure test will give you a straightforward 'yes' or 'no' answer about your health. Whole genome sequencing, though, will often give you a 'maybe.' Most of the information you get from a genomic test tells you about your risk for disease, not whether you have a disease or will for sure get the disease. Weather forecasting tries to predict what the weather will be like in an hour, a day, or a week. It involves knowledge of patterns and statistics. When a weather forecast predicts rain this coming weekend, it usually comes as a percentage, like 30% chance of showers.

During whole genome sequencing, researchers collect a DNA sample and then determine the identity of the 3 billion nucleotides that compose the human genome. The very first human genome was completed in 2003 as part of the Human Genome Project, which was formally started in 1990. The first human genome cost $2.7 billion. Today, most genetic testing focuses on one or a few genes, rather than the entire genome. However, with the falling cost of genome sequencing, more individuals are pursuing this option. Physicians can look at an entire genome to see how specific treatments for a disease will be affected by an individual's unique genetics. For example, the physician may opt to look at genes involved in drug metabolism when deciding dosage. In the future, whole genome sequencing may enable everyone to develop a personalized treatment plan.

It doesn't predict exactly what will happen, but gives you the chances of something happening. This means that it will tell you more about your risk for a certain disease, like diabetes, not if you have diabetes or not. But just because genomic information is often uncertain doesn't make it less useful. We can take that information and change our lifestyle choices to better our health.

9.2.1.1 Advantages

- ☆ Creating personalized plans to treat disease may be possible based not only on the mutant genes causing a disease, but also other genes in the patient's genome.
- ☆ Genotyping cancer cells and understanding what genes are misregulated allows physicians to select the best chemotherapy and potentially expose the patient to less toxic treatment since the therapy is tailored.
- ☆ Previously unknown genes may be identified as contributing to a disease state. Traditional genetic testing looks only at the common "troublemaker" genes.
- ☆ Lifestyle or environmental changes that can mediate the effects of genetic predisposition may be identified and then moderated.

9.2.1.2 Disadvantages

- ☆ The role of most of the genes in the human genome is still unknown or incompletely understood. Therefore, a lot of the "information" found in a human genome sequence is unusable at present.
- ☆ Most physicians are not trained in how to interpret genomic data.
- ☆ An individual's genome may contain information that they DON'T want to know. For example, a patient has genome sequencing performed to determine the most effective treatment plan for high cholesterol. In the process, researchers discover an unrelated allele that assures a terminal disease with no effective treatment.
- ☆ The volume of information contained in a genome sequence is vast. Policies and security measures to maintain the privacy and safety of this information are still new.

9.3 GENE EXPRESSION

A gene is a small piece of genetic material written in a code and called DNA. Each gene has within it a set of instructions for making molecules that organisms need to survive. Genes themselves cannot be used by an organism. Instead they must be turned into a gene product. Gene expression is the process by which the information contained within a gene becomes a useful product.

Gene expression is the phenotypic manifestation of a gene or genes by the processes of genetic transcription and genetic translation. Gene expression analysis is determination of the pattern of genes expressed at the level of genetic transcription, under specific circumstances or in a specific cell.

The translation of information encoded in a gene into protein or RNA structures that are present and operating in the cell. Expressed genes include genes that are transcribed into messenger RNA (mRNA) and then translated into protein, as well as genes that are transcribed into RNA, such as transfer and ribosomal RNAs, but not translated into protein.

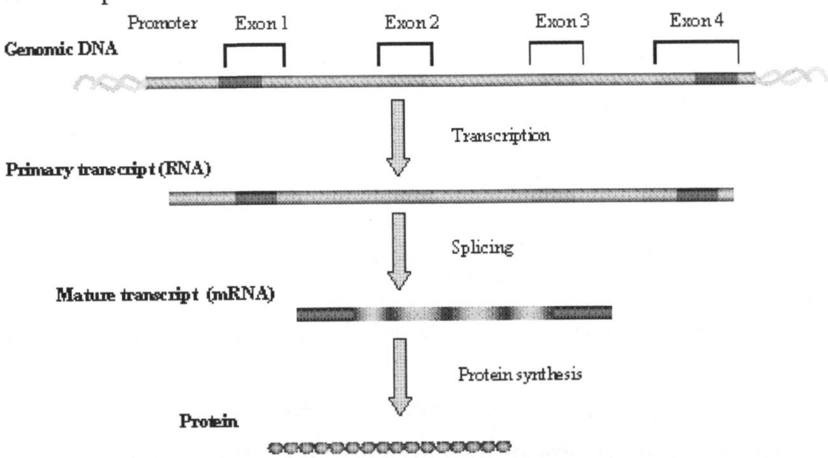

When genes are expressed, the genetic information (base sequence) on DNA is first copied to a molecule of mRNA (transcription). The mRNA molecules then leave the cell nucleus and enter the cytoplasm, where they participate in protein synthesis by specifying the particular amino acids that make up individual proteins (translation).

9.3.1 Regulation of Gene Expression

Genes can be expressed as either RNA or protein. However, not every gene product is needed all the time, nor are they needed in the same amounts. In fact, it would be very energetically demanding for a cell to express every gene all the time. Further, some gene products are harmful to the cell and thus only expressed at the end of a cell's life.

How do we know what gene product is needed, when it is needed, and how much is needed? Our cells actually have a very complicated system of mechanisms to monitor themselves and their environment. They take internal and external signals, analyze them, and then decide if a gene product is needed. Once it is decided that a gene product is needed, cells have a process to create a gene product.

The first step in making a gene product is to make a copy of the gene. Genes cannot leave the nucleus, so we need to make a copy, which is called messenger RNA (mRNA). The mRNA can leave the nucleus. So, how do we make mRNA? The first step is for a transcription factor to bind to the DNA at a point upstream to the target gene. This initiation point is a sequence of DNA called a promoter region.

Transcription is the process by which a transcript (mRNA) of the gene is created. Transcription factors can recruit the rest of the mRNA creating machinery to the target gene. The number and types of other factors varies with species, cell type, and even what the target gene is. An important component that is always present is RNA polymerase. This is the machine that actually reads the DNA code and makes the RNA.

CHAPTER-10

ENZYMES

10.1 DEFINITION

Enzymes are catalysts that accelerate the rate of chemical reactions without permanent alteration to themselves. Virtually all enzymes are proteins or conjugated proteins, although some catalytically active RNAs have been identified. These catalytic/enzymatic activities are essential to the information and energy management requirements of a cell. Specific enzymatic activities are found within all cellular organelles. In comparison with classical catalysts of chemical reactions enzymes are characterized by: (1) higher reaction rates; (2) effectiveness under milder reaction conditions in terms of temperature, pressure, and pH; (3) greater reaction specificity in terms of the reactants, products, and the absence of undesirable side reactions; and (4) ability to be regulated either by reaction rate control, by catalyst concentration, or by specific small molecules.

Enzymes are built of proteins folded into complicated shapes; they are present throughout the body. The chemical reactions that keep us alive - our metabolism - rely on the work that enzymes carry out. Enzymes speed up (catalyze) chemical reactions; in some cases, enzymes can make a chemical reaction millions of times faster than it would have been without it. A substrate binds to the active site of an enzyme and is converted into products. Once the products leave the active site, the enzyme is ready to attach to a new substrate and repeat the process.

Enzymes are made from amino acids, and they are proteins. When an enzyme is formed, it is made by stringing together between 100 and 1,000 amino acids in a very specific and unique order. The chain of amino acids then folds into a unique shape. That shape allows the enzyme to carry out specific chemical reactions -- an enzyme acts as a very efficient catalyst for a specific chemical reaction. The enzyme speeds that reaction up tremendously.

For example, the sugar maltose is made from two glucose molecules bonded together. The enzyme maltase is shaped in such a way that it can break the bond and free the two glucose pieces. The only thing maltase can do is break maltose molecules, but it can do that very rapidly and efficiently. Other types of enzymes can put atoms and molecules together. Breaking molecules apart and putting molecules together is what enzymes do, and there is a specific enzyme for each chemical reaction needed to make the cell work properly.

Enzymes have extremely interesting properties that make them little chemical-reaction machines. The purpose of an enzyme in a cell is to allow the cell to carry out chemical reactions very quickly. These reactions allow the cell to build things or take things apart as needed. This is how a cell grows and reproduces.

10.2 TYPES OF ENZYMES:

- ☆ **Exo-enzymes:** Enzymes that function outside the cell are called so, e.g. zymase, lysozyme, digestive enzymes.
- ☆ **Endo-enzyme:** Enzymes that function inside the cell are called so, e.g. enzymes of glycolysis, Krebs cycle, protein biosynthesis etc.
- ☆ **Zymogens:** These are inactive precursors or pro-enzymes forms of exo-enzymes. They become activated prior to enzymatic action, e.g., proteases.
- ☆ **Constitute or housekeeping enzymes:** Those enzymes are always present and synthesized in cell, e.g., glycolytic enzymes.
- ☆ **Inducible enzymes:** Most enzymes are synthesized only when they are needed e.g. Nitric oxide synthase, cycloxygenase, aldehyde dehydrogenase etc.
- ☆ **Isoenzymes (isozymes):** These are the different forms of the same enzymes which catalyze the same chemical reaction but, differ each other chemically, immunologically, and electrophoretically and in kinetic properties. For example, in maize 18 isozymes found for peroxidase. In plants aspertate kinase exist in two isozyme forms. Aspertate kinase catalyzes the amino acid biosynthesis from aspertate, LDH (Lactic acid dehydrogenase)
- ☆ **Ribozyme or RNA Enzymes:** e.g. ribonuclease-P (RNAase-P), Peptidyl transferase (23S rRNA of larger subunit of ribosome) etc.
- ☆ **Abzymes:** These are the antibodies that act as enzymes.

10.3 STRUCTURE

All enzymes are generally globular proteins except some RNA enzymes like Ribonuclease-P, ribozyme and peptidyl transferase. On the basis of number of polypeptide chains, enzymes are of 2 types:

- ☆ **Monomeric enzymes:** Consist of one polypeptide chain (subunit), e.g., ribonuclease, lysozyme, hexokinase etc. These are functional in their 3 dimensional or tertiary structures.
- ☆ **Oligomeric enzymes:** Consist of more than one oolypeptide chain. They are functional in their quaternary structure. For example, aldolase consists of 4 chains (tetrameric), Rubisco of Calvin cycle consists of 24 chains, and Enolase is a dimmer.

On the basis of chemical nature, enzymes are also 2 types:

- ☆ Simple enzymes: They consist of only proteins, e.g. urease, lysozyme, pepsin, trypsin etc.
- ☆ Holoenzyme or Conjugate enzyme: These enzymes consist of proteinous part called apoenzyme and non-proteinous part called co-factor.

Holoenzyme = Apcenzyme + Co-factor (active) (Proteinous part) (Non-proteinous part). The cofactors are of 3 types: Co-enzymes, Prosthetic groups and inorganic ions.

Catalytic Site or Active Site or Active Spot: It is a small three-dimensional (3D) area on or near enzyme surface that binds the specific substrate(s) and convert into products. The unique 3D shape of a catalytic site may alter by denaturation (unfolding) through high temperature or exposure to extremes of pH. This results in the loss of catalytic activity. In monomeric enzymes, the catalytic site is often a cleft or crevice, but in multimericenzymes, it resides at the interface between polypeptides. In some enzymes, the catalytic site is rigid or non- flexible to accommodate a substrate. But in most cases, the binding of substrate induces a conformational change in the catalytic site, e.g., glucose (substrate) induces a conformational change of hexokinase. An enzyme may have one or more active sites. Each active site consists of 3-12 amino acids that come together by folding of polypeptides. In a holoenzyme, the catalytic site also contains cofactor for its function. The amino acids residues of catalytic site lave 4 roles:

- ☆ Provide charged R-groups to attract substrate,
- ☆ Some act as template for holding substrate,
- ☆ Some provide functional groups that perform chemical changes by lowering activation energy

A few residues determine substrate specificity:

Enzymes : Amino Acid residues in active site

Pepsin : Tyrosine only

Aldolase : Glycine – Histidine – Alanine

Allosteric Sites: These are special sites on enzyme surface other than catalytic site, which when bind with effectors or modulators alter the conformation of the catalytic site. The enzymes having allosteric sites are called allosteric enzymes. Allosteric sites are of two types: activator site and inhibitor site. An allosteric activator when binds to activator site increase the enzyme activity while an allosteric inhibitor decreases the enzyme activity by binding the inhibitor site.

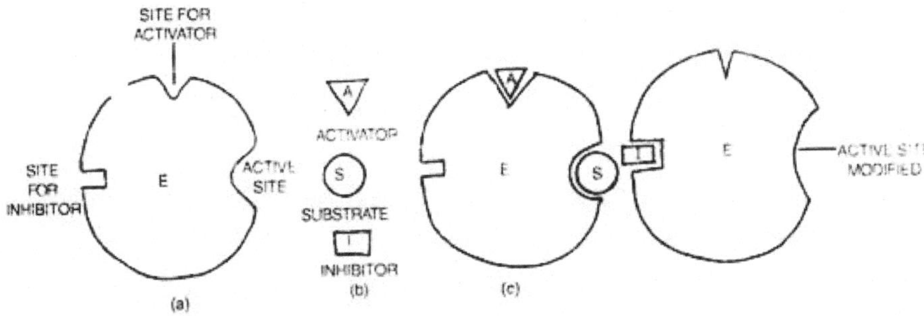

Fig Working of Allosteric Sites

10.4 CLASSIFICATION

Enzymes can be defined as the reaction catalysts of biological systems produced by living cells and are capable of catalyzing chemical reactions. Naturally enzymes are macromolecules having molecular masses ranging into millions. Two exceptional properties of enzymes are their extraordinary specificity — each enzyme catalyzes only one reaction or one group of closely related reactions — and their amazing efficiency — they may speed up reactions by factors of up to 1020. Each enzyme molecule possesses a region known as the active site and the substrate binds itself with this active site. Enzymes are either pure proteins or contain proteins as essential components and in addition require non-protein components which are also essential for their activity. The protein component of the enzyme is called apoenzyme and the non-protein component is called the co-factor or co-enzyme. The co-factors include inorganic ions and complex organic or metallo-organic molecules. Important inorganic co-factors along with their respective enzymes include Fe2+ (chrome oxidase) Ze2+ (carbonic anhydrase) and Mg2+ (glucose 6- phosphatase), etc. Many enzymes contain vitamins as their co-factors, for example; nicotinamide adenine dinucleotide contains nicotinamide vitamin and thiamine pyrophosphatase contains vitamin B1.

Enzymes can be classified in different ways as per:

- ☆ Location of enzyme
- ☆ Kind of substrate
- ☆ Enzyme reaction

Classification by the enzyme reaction is connected with the nomenclature of enzymes.

All reactions in the body take place as a result of the actions of enzymes. Therefore, if an enzyme is localized where a metabolic response is involved, the enzyme exists universally and is classified into membrane enzymes bound with the biomembrane (cell membranes and the membranes of cell organelles) and soluble enzymes that exist in the cytoplasm or luminal fluid. Some soluble enzymes are secreted out of cells. Differences between enzymes are based on the difference in location on the structures of the hydrophobic and hydrophilic side-chains, based mainly on the amino acid sequence.

International Union of Biochemistry (IUB, 1961) adopted a scheme for systematic functional classification and nomenclature of enzymes. The recommendations of IUB are as follows:

- ☆ All known enzymes have been grouped into six major classes on the basis of reaction type they catalyze,
- ☆ Each class further sub-divided into subclasses and sub-subclasses,
- ☆ Each enzyme is assigned two names i.e., recommended (trivial) name and systematic name,
- ☆ Each enzyme is identified by a unique four digit classification number.

For example, hexokinase is recommended name, its systematic name is glucose phosphotransferase and its classification number in EC 2.7.1.1. Here, "EC" stands for Enzyme commission, the first number (2) stands for the major class, the second number (7) stands for the sub class, the third number (1) indicates sub-class and the fourth number (1) denotes the serial number assigned in its sub-classes.

Table: IUB Classification

Major Class (Type of reaction catalyzed)	Common exmaples	Kind of reaction	Specific Example
1. ***Oxidoreductases*** (*Transfer of electrons*)	Oxidases Reductases Dehydrogenase	$A^{+3} + B^{+2} \rightarrow A^{+2} + B^{+3}$	Alcohol + NAD ↓ *Alcohol dehydrogenase* Aldehyde + $NADH_2$
2. ***Transferases*** (*Transfer of functional groups*)	Transaminase Transketolase Transaldolase	A – X + B → A + B – X	Glucose + ATP ↓ *Glukokinase or hexokinase* Glucose-6-Phosphate + ADP
3. ***Hydrolases*** (*Hydrolysis Reactions*)	Amylases Lipases Proteases Nucleases	A–B+H_2O→A–OH + B–H	Sucrose ↓ *Sucrase* Glucose + Fructose
4. **Lyases or *Desmolases*** (*Group elimination to form double bonds without hydrolysis*)	Aldolase Decarboxylase Fumarase Citrate synthase	A – B → A = B + X – Y I I X Y	Histidine ↓ *Histidine decarboxylase* Histidine + CO_2
5. ***Isomerases*** (*Transfer of Groups within a molecule*	Isomerase Mutase Epimerase	A – B → A – B I I I I Y X X Y	Glucose – 6-Phosphate ↓ *Isomerase* Fructose-6-Phosphate
6. ***Ligases or Synthetases*** (*Bond formation couples with ATP hydrolysis*)	Synthetases Carboxylases	A + B + ATP → A – B + ADP + Pi	Pyruvate + CO_2 + ATP ↓ *Pyruvate carboxylase* Oxaloacetate + ADP + Pi

Although enzymes, as well as other proteins, are biosynthesized by intracellular ribosomes, the evolution of enzymes may be reflected in the structure because the motif is dependent on a particular gene. Hereditarily, similar enzymes have a similar motif and form an enzyme group.

10.5 MEMBRANE ENZYMES

Many enzymes in the biomembrane participate in the conservation of energy or transportation of substances making up important enzyme groups (5′-nucleotidase, γ-glutamyltransferase, etc.). The membrane enzymes can be divided into three kinds according to the spatial relationship of the biomembrane and the enzyme. The first type, integral membrane enzymes, is buried in the biomembrane (receptor proteins, etc.); the second penetrates the biomembrane (channel transporter, etc.); and in the third type, peripheral membrane enzymes, the enzyme is attached to the biomembrane (alkaline phosphatase, γ-glutamyltransferase, etc.). Since the biomembrane has a hydrophilic surface and an inner hydrophobic fraction (lipid bilayer membrane), the membrane anchoring moiety (hydrophobic peptide chain) of the enzyme links with the membrane by strong hydrophobicity. Regarding the affinity to the membrane, the lipid moiety also has very strong affinity to the membrane as a glycolipid moiety (GPI anchor).

10.6 SOLUBLE ENZYMES

Cytosolic enzymes exist in their solubilized form in the liquid phase in cells or luminal fluid in the bloodstream. Many of these soluble enzymes are concerned with metabolism in the cytoplasm. Many hydrophilic amino acids are positioned on the outside of soluble enzymes and hydrophobic amino acids are found on the inside of the structure.

Although enzymes are produced within certain cells, some enzymes, called secretory enzymes, are secreted out to the cytosol fraction after formation. For example, digestive enzymes digest substances to facilitate their absorption into the cells. Many secretory enzymes are excreted in soluble forms. If a certain stimulus is applied to the body, secretory enzymes are released through exocytosis. Research in structural biology has identified the 3D structure of soluble enzymes by crystallization techniques.

GLOSSARY

Active site: The molecular location in the small intestine where enzymes and substrates come together.

Allele: One of two or more alternative forms of a gene that arise by mutation and are found at the same place on a chromosome.

Amylase: A digestive enzyme that breaks down carbohydrates such as starch and glycogen into smaller units; there are 2 types of amylase- salivary and pancreatic.

Amylopectin: The linear form of starch that consists of α- (1,4) glucose polymers as well as branched α- (1,6) glucose polymers; amylopectin starches are digested more rapidly than amylose starches. Amylopectin makes up about 70-80% of the structure of starch.

Amylose: The linear form of starch that consists of α- (1,4) linkages of glucose polymers; amylose starches are less rapidly digested than amylopectin starches. Amylose makes up about 20-30% of the structure of starch.

Anus: A muscular valve that controls the opening from the large intestine to facilitate waste elimination (bowel movements).

Autosomal recessive: The pattern of inheritance of CSID; a mutation in a gene on one of the first 22 non-sex chromosomes that can lead to an autosomal disorder such as CSID. Recessive inheritance means both genes in the pair must be defective to cause disease.

Biopsy: A tissue sample; in digestive diseases, the biopsy is usually painless and is taken from the inner layers of the esophagus, stomach, small, and/or large intestines during endoscopy and colonoscopy procedures.

Bloating: A fullness or distention of the abdomen, usually caused by excess gas that accumulates in the small and/or large intestines. CSID can cause bloating; some foods such as beans, peanuts, cabbage, cauliflower, broccoli, and kale may lead to bloating from fermentation due to a high level of a certain carbohydrate in these foods. Also, improper digestion of other nutrients (fats, carbohydrates, proteins) may cause bloating.

Brush border hydrolases: Enzymes present in the brush border or microvilli of the small intestine, which break down disaccharides and starches; examples are sucrase, isomaltase, and lactase.

Carbohydrate: Any substance that has a molecular structure based on sugar, including sugar, glucose, fructose, etc., as well as more complex starches found in plant food sources (fruits, grains, vegetables, etc.); all starches are carbohydrates, but not all carbohydrates are starches (i.e., honey and grapes are carbohydrates, but not starches).

Colonoscopy: A procedure in which a camera on the end of a lighted tube, an endoscope, is inserted through the anus and rectum and is used to view the inside of the entire large intestine and sometimes the terminal ileum. Biopsies of the mucosal layers within the large intestine or terminal ileum of the small intestine may be taken during the colonoscopy procedure. Preparation involves the use of medications to clear out waste from the large intestine and avoiding food and drink for at least 12 hours before the procedure. This procedure requires sedation.

Compound heterozygote: A person with CSID who has two different mutated alleles at a particular location on a gene, one on each chromosome 3 of a pair; the human genome contains two copies of each gene, 1 paternal and 1 maternal allele. A mutation affecting only one allele is called heterozygous. A person is homozygous if they have two identical forms of a particular gene, one inherited from each parent. However, when a person inherits a mutated maternal allele and a different mutated paternal allele, these mutations are called compound heterozygous.

Congenital: A medical term which means present at birth.

Disaccharides: Sugars that require brush border enzymes for digestion; sucrose, lactose, and maltose are all disaccharides.

Duodenum: The upper third of the small intestine located between the lower part of the stomach and the jejunum.

Dyspepsia: A medical term for indigestion or upset stomach that describes discomfort or pain in the upper abdomen; dyspepsia is not a disease but rather a group of symptoms which often include bloating, nausea, and burping.

egistered dietitian: Registered dietitians (RDs) are experts in food and nutrition who have met both academic and professional requirements. RDs can be extremely beneficial for a patient with CSID.

Endoscope: A tube with a light and a camera on the end of it that is used to view the esophagus, stomach, part of the duodenum, part of the terminal ileum, and the entire large intestine. Endoscopes usually contain a biopsy aspiration port to collect tissue samples during endoscopic procedures.

Enzyme: A type of catalyst, which means it speeds up chemical reactions. In the case of digestive enzymes, they speed up the digestion or breaking down of food into nutritional components such as fatty acids, amino acids, sugars, vitamins, and minerals. Some metabolic disorders cause a decrease of a certain enzyme needed for processing or metabolizing specific food components, such as the case with sucrose and CSID.

Failure to Thrive (FTT): A general term used for insufficient growth (by standard growth charts) or inadequate weight gain caused by a variety of medical and psychosocial conditions; does not imply abnormal intellectual, social, or emotional development.

Flatulence: Passing gas through the anus; may be accompanied by bloating, indigestion, and/or other symptoms.

Fructose: A monosaccharide, simple sugar found in many plants; 1 of 3 monosaccharides (others are glucose and galactose) that are absorbed directly into the bloodstream during digestion.

Galactose: A monosaccharide, simple sugar; galactose and glucose are the monosaccharides from the breakdown of lactose.

Gastroenterologist: A physician who specializes in disorders of the gastrointestinal tract and associated organs.

Genetics: The branch of biology that studies heredity and variation in organisms.

Genotype: The inherited instructions a person carries within his/her genetic code or DNA.

Glucoamylase: An enzyme that hydrolyzes the glucoside bond in starches and dextrins; breaks down carbohydrates, specifically polysaccharides into glucose.

Glucose: A monosaccharide, simple sugar that is the building block of starch and maltase and a result of broken down lactose or sucrose; glucose is needed by all cells for energy.

Glycosidases: Enzymes that hydrolyze glycosidic bonds such as those seen in disaccharides like sucrose.

Glycosidic bonds: A glycosidic bond is a type of bond that joins a sugar (carbohydrate) molecule to another group, which may or may not be another sugar (carbohydrate); the type of chemical linkage between the monosaccharide units of disaccharides, oligosaccharides, and polysaccharides.

Heredity: The passing of traits from parents to offspring, also called genetics.

Heterozygote: A person who has two different forms of a particular gene; a heterozygote for CSID has a mutated gene for the disorder from one parent and a normal gene from the other parent; also known as a carrier.

Homozygote: A person with CSID who has two identical mutated forms of a particular gene, one inherited from each parent.

Hydrolysis: The process by which a molecule is broken down by adding water; hydrolysis is a step to break down the glycosidic bonds of disaccharides.

Ileum: The last third of the small intestine, which connects the jejunum with the large intestine.

Isomaltase: The enzyme partially responsible for the breakdown of starches; maltase-glucoamylase also plays a part in starch digestion; isomaltase is synonymous with palatinase; isomaltase breaks the bonds linking saccharides, which cannot be broken down by amylase and maltase.

Jejunum: The middle third of the small intestine.

Lactase: Brush border enzyme that breaks down lactose, the sugar found in milk; some CSID patients are deficient in lactase, but most are not; many CSID patients have higher than normal lactase levels to compensate for the lack of SI activity.

Lactose: Disaccharide found in milk that is broken down by the enzyme, lactase; made up of 1 glucose unit and 1 galactose unit.

Lactose intolerance: A deficiency in the intestinal enzyme lactase, which does not allow for lactose to digest or absorb properly. Malabsorbed lactose reaches the large intestine and is fermented by bacteria. Byproducts of the fermentation process cause symptoms of flatulence, abdominal cramps, and/or diarrhea after the ingestion of lactose (milk sugar).

Large intestine (large bowel): A long muscular tube that connects the small intestine with the anus; primary function is to absorb water from waste before it leaves the body.

Lumen: The hollow portion of a tubular body structure, such as a blood vessel, the intestines, and the esophagus.

Malabsorption: A clinical term that describes the inability to absorb sufficient nutrients, either certain specific nutrients or all nutrients in general, from one's typical diet.

Maltase-glucoamylase: A brush border membrane enzyme that plays a role in the final steps of small intestinal digestion of starch to glucose.

Maltase: An enzyme that breaks down the disaccharide maltose. Maltose is present in many starchy foods.

Maltose: A disaccharide that requires maltase for digestion; breaks down into 2 molecules of glucose in digestion; maltose is usually found in grains and starchy vegetables.

Metabolic disorder: A disorder in which the production or breakdown of one or more specific chemical substance(s) within the body is abnormal due to a genetic alteration in a single enzyme.

MGAM gene: The gene responsible for the expression of maltase-glucoamylase in the body; works closely with the SI gene to digest starch.

Microvilli (villi): Tiny, finger-like projections that protrude from the lining of the small intestinal wall; serve to increase the internal surface area of the intestinal walls which assists with nutrient absorption.

Monosaccharides: A simple sugar composed of only one sugar molecule; glucose, galactose, and fructose are monosaccharides.

Motility: The coordinated neuromuscular activity of the wall of the GI tract that moves intestinal contents.

Oligosaccharides: Sugars made up of 3-10 sugar units.

Orphan disease: A disorder that affects a small population of patients. In the United States, an orphan disease is defined by law as having a prevalence of less than 200,000 individuals. In Europe, an orphan disease is defined as diseases affecting less than 5 individuals per population of 10,000.

Orphan drug: A drug that is used to treat a small population of patients. In the United States, an orphan drug treats an orphan disease. (see orphan disease)

Osmotic diarrhea: Diarrhea that is caused by the presence of unabsorbed nutrients in the gastrointestinal tract, leading to impairment in digestion or absorption. In CSID, sucrose is not enzymatically cleaved and absorbed in the small intestine so it reaches the colon in an intact form, where the colonic bacteria ferment it, which drives water into the colon leading to diarrhea.

Palatinase: A brush border enzyme that hydrolyzes palatinose; sometimes used interchangeably with isomaltase; usually palatinase is measured during a disaccharidases test in a small bowel biopsy.

Phenotype: The observable physical and/or biochemical characteristics of an organism, as determined by both genetic makeup and environmental influences; the expression of a specific trait based on genetic and environmental influences.

Polymer: A chemical compound or mixture of compounds consisting of repeating structural units.

Polysaccharides: Sugars made up of more than 10 sugar units; starch is a type of polysaccharide.

Prevalence: The total number of persons in a given population with a disease or other health-related event during a specified period of time, usually expressed as a percentage.

Ptyalin: A term sometimes used interchangeably with salivary amylase.

SI gene: The gene responsible for the expression of sucrase-isomaltase in the body.

Sigmoidoscope: A short, flexible, or rigid endoscope designed only to examine the rectum and sigmoid portion of the large intestine.

Sigmoidoscopy: A procedure in which a flexible or rigid sigmoidoscope is inserted into the rectum and through the entire length of the sigmoid portion of the large intestine for examination and possible biopsy. Preparation includes an enema and mild laxative to clear waste from the sigmoid and rectal areas. Sedation may be used if needed.

Small intestine (small bowel): A long, muscular tube that connects the bottom of the stomach with the large intestine (large bowel); primary function is to complete digestion and nutrient absorption into blood stream.

Starch: A complex carbohydrate found chiefly in seeds, fruits, tubers, and roots and stem pith of plants, notably in corn, potatoes, wheat, and rice; all starches are carbohydrates, but not all carbohydrates are starches (i.e., honey and grapes are carbohydrates, but not starches, while bread is both a carbohydrate and a starch).

Substrate: A molecule that is affected by an enzyme(s); sucrose is the substrate for sucrase.

Sucrase: The brush border enzyme that is responsible for breaking down sucrose, a disaccharide, into glucose for the body to use as fuel; patients with CSID have little to no sucrase activity.

Sucrose: A disaccharide that requires brush border enzymes for digestion, primarily sucrase; breaks down into fructose and glucose in digestion.

CHAPTER- 11

VITAMINS AND COENZYMES

11.1 INTRODUCTION

Vitamins are a multibillion dollar industry.1 They are readily available to the public and are the focus of the most frequently asked questions to pharmacists.2 Surprisingly, most pharmacy students receive little training on the many roles of vitamins in nutrition.2 Therefore, this lecture attempts to not only reinforce fundamental biochemical and metabolic pathways in the human body, but also to provide pharmacy students with practical information.

The diet is a vast source of important nutrients. These nutrients include several important classes of biomolecules such as: (1) energy yielding components (carbohydrates, lipids, and proteins), (2) essential and nonessential amino acids, (3) essential fatty acids, (4) minerals, and (5) vitamins. Vitamins are organic substances that must be provided by the diet either because they cannot be biosynthesized or the amount that is provided through biosynthesis is inadequate for maintaining normal health.

11.2 CLASSES OF VITAMINS

Vitamins are broadly divided into 2 classes based upon their hydrophobicity. The more hydrophilic vitamins are termed the water-soluble vitamins and are composed of the B-complex vitamins and vitamin C. The more hydrophobic vitamins, referred to as the fat-soluble vitamins, are composed of Vitamins A, D, E, and K.

The normal North American diet is sufficient to prevent significant vitamin deficiencies and the related diseases associated with these deficiencies. However, there is increasing concern that slight vitamin deficiencies in a number of water-soluble vitamins (B1, B6, B12, folate and vitamin C) are risk factors for diseases such as such as depression, cancer, cardiovascular disease, and neural tube defects. The prevalence of this slight vitamin deficiency is likely related to diet, since most of these vitamins are supplied by fruit and vegetable intake. A recent survey has shown that only 25% of the population meets their daily intake of 5 servings.3 Food preparation is another source of vitamin loss. For example, heating food for more than 2 hours causes more than a 10% loss of most water-soluble vitamins. Refrigeration, freezing, and reheating have all been shown to lead to further loss of vitamins. Exposure to light causes significant loss of riboflavin from foods and the combination of heating and light can almost completely remove this vitamin from food.

11.2.1 Thiamin (Vitamin B1)

Thiamin functions in the body as thiamin pyrophosphate (TPP) an important enzyme co-factor.5 Thiamin reacts with adenosine triphosphate (ATP) to form thiamin pyrophosphate through a reaction mediated by the enzyme thiamin diphosphokinase. Following its production, TPP is incorporated into 2 important enzymes: pyruvate dehydrogenase and ketoglutarate dehydrogenase. Pyruvate dehydrogenase is part of a multi-enzyme complex that acts to convert pyruvate generated in glycolysis into acetyl-CoA for entry into the tricarboxylic acid (TCA) cycle. Thiamin pyrophosphate is also used as a co-factor for the enzyme -ketoglutarate dehydrogenase, which is a key point of regulation in the TCA cycle. -Ketoglutarate dehydrogenase is involved in the conversion of ketoglutarate to succinyl CoA. In the case of both enzymes, TPP assists in decarboxylation of a small ketoacid.

In addition to its uses in metabolism, thiamin may enhance circulation and blood formation. It is required for maintenance of the nervous system and is used in the biosynthesis of the neurotransmitters acetylcholinc and *hydroxybutyrate (GABA). Thiamin also is used in the production of hydrochloric acid in the stomach and, therefore, has a role in digestion.

Thiamin is provided in the diet through most grains. A deficiency of thiamin is called beriberi and it is most often observed in Southeast Asia.6 The main staple of diet in this part of the world is rice, which does not provide dietary thiamin to as great an extent as other grains. The symptoms of beriberi include limb pain, muscle weakness, and low cardiac output. all symptoms are related to the diminished capacity of the major energy producing pathways that are dependent on TPP as a co-factor.

Mild deficiencies of thiamin are sometimes observed in the elderly and in low-income groups on restricted diets. The earliest symptoms of thiamin deficiency are loss of appetite, constipation, and nausea. Other symptoms such as mental depression, peripheral neuropathy, irritability, and fatigue are related to the role of thiamin in maintaining a healthy nervous system.

The most common form of thiamin deficiency in the United States is alcoholic neuritis. Because alcoholics normally have a poor appetite, overall food consumption is low. In addition, alcoholics are predisposed to developing nutritional deficiencies since alcohol is their primary calorie source. Under the broad umbrella of alcoholic neuritis come two disorders, Wernicke's Syndrome and Korsakoff s Psychosis.7 The symptoms of Wernicke's Syndrome (ophthalmoplegia, nystagmus, and ataxia) respond quickly to administration of thiamin, whereas the more severe memory and learning disorders associated with Korsakoff s Psychosis are refractory to thiamin treatment.

11.2.2 Riboflavin (Vitamin B2)

The reduced forms of these enzyme co-factors can donate these electrons to return to their previous fully oxidized forms. It is this ability to act as a conduit for electron transfer reactions that makes FAD and FMN such important enzyme cofactors.

Riboflavin is available from a wide variety of dietary sources such as milk, cheese, meat, eggs, and cereal products. Symptoms associated with riboflavin deficiency include sore throat, dermatitis, anemia, neuropathy, and cataract formation. Riboflavin deficiency is not normally observed in the United States; however, deficiency may be observed as part of a general case of malnutrition or in cases of chronic alcoholism. The conversion of riboflavin to FMN is required for absorption and transport into many tissues. The conversion has been shown to be inhibited by hypothyroidism and the structurally similar medications chlorpromazine, adriamycin, quinacrine, tetracycline, and tricyclic antidepressants.

11.2.3 Niacin (Vitamin B3)

Niacin can be found in foods such as meats, breads, and beans. Mild niacin deficiencies have similar symptoms to those observed with riboflavin, which is not surprising due to the similar roles both of these vitamins play in biochemical reactions. Niacin deficiencies are occasionally observed in alcoholics, cases of general malnutrition, and in the elderly on restricted diets. A severe deficiency of niacin is known as pellagra, which is derived from the Italian phrase meaning rough skin. Pellagra is marked by dermatitis and also is notable for causing a blackening of the tongue. Early cases of pellagra were first observed in Europe shortly after the introduction of corn from the voyages of Christopher Columbus. In these cases, poor farmers who were raising corn as animal feed were particularly susceptible. Because niacin in corn is not bioavailable unless treated with a strong base such as lye, the farmers developed niacin deficiencies. Europeans did not know this processing method until revealed to them by Native-Americans during the early colonial period of North America. However, the connection between pellagra and niacin was not known until the early 20th century. Pellagra was a significant health issue in the United States over the period from 1900-1940 resulting in over 100,000 deaths. Today most diets are supplemented with niacin through enriched flour, which receives its name because of the added niacm.

Niacin can be administered in doses of 2g to 4g to causes a decrease in circulating levels of cholesterol and LDL. While the cholesterol lowering effects of niacin are desirable there are potential side effects from such large doses of this vitamin. The most immediate reaction observed from large doses of niacin is vasodilation resulting in flushing. Over time there may be a reduction in fatty acid mobilization causing a depletion of glycogen and lipid stores in muscle tissue. Long-term exposure may also elevate blood glucose and uric acid levels, suggesting increased risk for patients who are on the borderline for diseases such as diabetes and gout. Prolonged use of high doses of niacin can lead to elevated levels of the serum enzymes alanine aminotransferase and aspartate aminotransferase, which may suggest liver damage.

11.2.4 Pyridoxine (Vitamin B6)

Pyridoxine is the precursor to the active enzyme cofactor pyridoxal phosphate (PLP). Pyridoxal phosphate is a critical co-factor for enzymes involved in reactions involving many amino acids. The N-terminus of the amino acid forms a covalcnt bond to PLP, allowing a wide variety of displacement reactions to occur at the alpha carbon. These include decarboxylations, transaminations, and transfers of side chains. Pyridoxine, therefore, plays a central role in the production of many neurotransmitters, such as serotonin, norepinephrine, and histamine. Pyridoxine also is important in the production of heme.

The enzyme LAAAD is present in both the gut and in the brain. This presents a problem because L-dopa is readily transported across the blood brain barrier while dopamine is not. Therefore, orally administered L-dopa is rendered ineffective in the gut by LAAAD. Co-administration of carbidopa resolves this dilemma by inactivating LAAAD in the gut. Like dopamine, carbidopa cannot cross the blood brain barrier leaving LAAAD in the brain free to carry out the production of dopamine at the site of action. In addition to hydrazine containing drugs, penicillamine, used in the treatment of Wilson's disease, cystinuria, and rheumatoid arthritis reacts with and inactivates pyridoxal phosphate. Patients treated with penicillamine occasionally develop convulsions, which can be prevented by supplementation with vitamin B6.

Pyridoxine is found in foods such as meats, breads, eggs, soybeans, and many vegetables. A deficiency in pyridoxine can cause facial lesions, depression, peripheral neuropathy, and glossitis. The neurological complications can be directly linked to the effects on neurotransmitter production. Mild pyridoxine deficiencies are sometimes observed in young women taking oral contraceptives. Deficiencies have also been observed in patients with gasteroenteritis or Crohn's Disease, presumably due to poor absorption. Mild deficiencies are of concern due to a correlation with an increased incidence of breast cancer and to the risk of coronary heart disease. Neurotoxicity has been noted with doses of vitamin B6 in excess of 500 mg/day.

11.2.5 Cobalamin (Vitamin B12)

Cobalamin is vital for cell growth and replication. Its major site of action is at the interface between the folic acid cycle and the active methyl cycle, where cobalamin is a co-factor for the enzyme homocysteine methyltransferase. This enzyme catalyzes the transfer of a methyl group from tetrahydrofolate to homocysteine, forming the amino acid methionine. These 2 cycles impact many other pathways due to the large number of memylation reactions in the body, especially nucleic acid biosynthesis and neurotransmitter biosynthesis.

Cobalamin is found in meats and dairy products. Deficiencies are not common because the liver can store a 6-year supply of cobalamin. In addition, cobalamin is highly conserved by enterohepatic recirculation. Strict vegetarians (vegans) may take up to 20 to 30 years to develop a deficiency, whereas inadequate absorption from the ileum due to ileitis or loss of a glycoprotein that complexes with cobalamin prior to absorption may take from 2 to 10 years to become symptomatic. The anesthetic nitrous oxide inactivates cobalamin and can cause patients with marginal serum levels to develop deficiencies within a week.

While rare, deficiencies in vitamin B12 are severe and manifest in erythrocytes and nervous tissue. In erythrocytes, nucleic acid biosynthesis is slowed. This results in stem cells that are committed to development into erythrocytes possessing twice their normal protein content but lacking the necessary nuclear material for proper cell division. The resulting anemia, termed megaloblastic anemia, can be quite severe. Cobalamin deficiency can also cause permanent damage to the nervous system. The results are swelling of neurons and demylination of nerve cells, followed by cell death. These progressive manifestations cause a wide range of neurological symptoms including unsteadiness, decreased reflexes, paresthesias of the extremities, and ultimately confusion, memory loss, hallucinations, and psychosis. The neurological symptoms of cobalamin deficiency may be mistaken for multiple sclerosis. In the elderly and in alcoholics, cobalamin deficiency should be considered as a possible cause of dementia even in the absence of anemia. The neurological symptoms associated with cobalamin deficiency are not directly related to irregularities in the formation of erythrocytes.

Nutrient depletion or shortage may arise from inadequate food intake, impaired digestion, inefficient absorption or malabsorption, inefficient assimilation and other physiologic or functional disorders. Genetic errors may interfere with or prohibit assimilation and utilization. Gastrointestinal motility disorders may have similar effects. Stress may affect both gastrointestinal tract transit times and quantities or ratios of nutrient needs, leading to depleted nutrients. Americans now spend $2.5 billion yearly on vitamins and minerals, up from $328 million 25 years ago. Roughly 55% of women and 45% of men take some kind of vitamin supplement. Most of the vitamin supplements consumed by Americans, including those used in "enriched" foods, are making them more deficient! The greatest and most enduring of medical discoveries of the twentieth century were the vitamins. Here, at last, was the discovery of factors that caused health. Vitamin A (1912, the first vitamin to be discovered) was named retinol because, without it, a healthy retina in the eye could not be formed. Then it was found to cure night blindness. And that was just the beginning.

Along came the B-vitamins and the cures for beriberi, pellagra, pernicious anemia, nerve degeneration, enlarged heart, energy production, and much more. These came from unprocessed, whole, natural foods. When man realized his foods had been devitalized from processing and refinement, he decided to replace the lost vitamins. But, in his effort to extricate himself from the refined food trap, "scientific confusion" led him to try to replace them with "chemical, man-made vitamins." So, again, he refined down natural vitamins into pure, crystalline vitamins, thus robbing them of important food values, just as he had refined natural grains and sweets into their pure, crystalline, carbon-white flour and sugar. Then, to save money, he duplicated these crystalline vitamins synthetically, mostly from COAL-TAR products. Thus, natural vitamins as well as natural foods were robbed of nature's life and valuable elusive nutrients.

The latest round in conventional medicine's ongoing attempts to discredit (and ultimately outlaw) nutritional supplements is found in a highly questionable study published February, 2007 in the Journal of the American Medical Association, which claims that vitamins actually increase the risk of death.

The study claims to have analyzed a collection of previous studies on Vitamin A, beta carotene, Vitamin C, Vitamin E and selenium, concluding that most of the nutrients are actually dangerous to human health. Of course, this is research from conventional medicine – an industry that promotes patented chemicals as perfectly safe, even though FDA-approved pharmaceuticals are killing at least 100,000 Americans each year. (Imagine the uproar if vitamins killed even a fraction of that number)

To avoid getting hoodwinked by questionable research on "vitamins," you have to strongly consider the financial interests of the source of this research. JAMA accepts millions of dollars in advertising from drug companies each year, and its pages are absolutely packed with drug ads. The American Medical Association, for its part, has long worked to discredit alternative medicine and has even been found guilty by U.S. federal courts of engaging in a conspiracy to destroy chiropractic medicine. The AMA, which is largely considered a joke by anyone familiar with natural health, is hardly a credible source for publishing scientific findings on nutrition. To protect the multi-billion dollar drug industry, it's likely the AMA would say practically anything.

This research published in JAMA does remind us of one important point, however: synthetic chemicals are harmful to human health. If you take cheap "vitamins" made of these synthetic chemicals, you are doing yourself more harm than good. These cheap vitamin manufacturers, by the way, are usually owned by pharmaceutical firms.

Conventional medicine researchers try to blur the line between "junk vitamins" and "quality vitamins" by classifying ALL nutritional supplements as "vitamins," regardless of what they're really made from. By discrediting a few synthetic chemicals, they can effectively dissuade the masses from taking ANY vitamins, including the good ones. And that is, of course, their goal: to use F.U.D. (fear, uncertainty and doubt) to scare consumers away from nutritional supplements so that patients will flock to the patented, synthetic chemicals that earn drug companies billions of dollars in profits. Drugs make money for Big Pharma, and vitamins compete with drug sales.

Round up 100 people who are taking multiple pharmaceuticals, and compare their health to 100 people who are taking vitamins and nutritional supplements. Guess who's healthier? The supplement crowd will be healthier every time. It's the obvious question: If vitamins are so dangerous, where are all the dead vitamin takers? And if pharmaceuticals are so safe, where are all the super-healthy prescription drug patients? They are nowhere to be found.

11.3 DESCRIPTION

It is a complex of biological "wheels within wheels" consisting of enzymes, coenzymes, antioxidants and trace mineral activators. Enzymes, being protein, contain amino acids and trace minerals. Every mineral and amino acid needed by living cells may be found in a natural assemblage of vitamin concentrates obtained from wholesome foods. Identifying a vitamin in terms of a single chemical structure is self-defeating--you need the whole complex to get the vitamin function. The natural vitamin complex carries trace mineral activators and without them, the vitamin fails in its metabolic function. Copper in its organic, biologically active form tyrosinase, is the trace element activator of the vitamin C-complex. Likewise, manganese activates the B complex. In turn, selenium is the trace mineral activator of the vitamin E-complex. Enzyme activators, destroyed by heat, steam sterilization and pasteurization, include Manganese, Cobalt, Molybdenum, Zinc, Selenium and Vanadium...and are effective only in the organic state. By reducing a vitamin complex to a single chemical, chemists can claim to duplicate it in a laboratory and easily mass produce it and sell it as an additive to food or a pill. This wouldn't be possible if they were charged with keeping all the synergistic components intact. As such, vitamins are given single chemical names and single chemical structures for the purpose of selling synthetic vitamins.

The failure of the pharmaceutical approach to vitamins originated in the chemists' synthetic isolation of elements, uncombined with its biological matrix. The discovery of the vitamins coincided with the boom in industry, agriculture, and chemistry, which were crowding their way into all facets of life. But in the attempted duplication of living things, like artificial fertilizers or vitamins, chemistry had its most influence. The biological activity of synthetic and natural vitamins is not identical. Anyone who has studied anatomy, physiology, neurology and biochemistry will readily admit that the human body is a neuro-biochemical plant--a composite of tissues, fluids and organs that exist in a living state by virtue of neuroelectric and biochemical reactions. Being biochemical in nature, the body operates on biochemicals--LIVE CHEMICALS--in contrast to dead or lifeless chemicals. Those biochemical factors are derived from whole, natural, uncooked and unprocessed foods. Cooking and processing alters biochemicals to dead chemicals. The human body cannot properly function if it is fueled with dead chemicals rather than live, complex biochemicals. The difference between the live biochemical food factor and the dead chemical (counterfeit food) is dependent upon two primary requisites: (1) unaltered proteins and (2) viable enzyme systems.

Synthetic vitamins are reverses conformationally or mirror images of natural vitamins. However, synthetic vitamins are not vitamins at all but synthesized fractions of a vitamin complex...a mirror image duplicate of just a portion of the real, biologically active and physiologically-precise, nutritional complex. There is no way that a fraction of a vitamin can be called a vitamin." As seen, vitamin B1 was first isolated in year 1926 and 10 years later, scientists were able to synthesize the mirror image of Thiamine from coal-tar. About the beginning of WWII enrichment of refined, devitalized flour became compulsory. Bleached wheat flour has more than 30 known nutrients removed with 4 synthetics added back in. Today, the majority of people in the civilized world are becoming victims of degenerative disease resulting from chronic poisoning and malnutrition. Even though well fed, they are starving to death for the most essential micronutrients. Since World War II, the American people and people of other countries as well, have had a daily ration of a genetic poison in most of the bread, flour products, cereals and other food items that are forced, by law, to be enriched with synthetic vitamins.

As observed, objective of finding the best ways to create foods that are non-perishable; can be made by mass-production methods, in centrally located factories and distributed so cheaply that they can sweep all local competition from the market. Then, after there develops a suspicion that these 'foods' are inadequate to support life, modern advertising science steps in to propagandize the people into believing that there is nothing wrong with counterfeit or devitalized foods, that they are products of value...and...the confused public is totally unable to arrive at any conclusion of fact. In turn, they continue to buy the rubbish that is killing them off, years prematurely."

In 1941, food research scientist reported that animals on a synthetic-vitamin-enriched diet: "dropped dead long before the animals on an un-enriched diet became disabled." She remarked at that time, that such phony enrichment might "precipitate conditions worse than the original deficiency." As per discussion on optical isomers or asymmetrical molecules as seen, atomic arrangements are mirror images of each other states as its structure is reversed, a left-handed molecule can't take part in chemical reactions meant for a right-handed molecule any more than a left hand can fit in a right-handed glove--its odd geometry prevents it from being metabolized by the body. In turn, the human physiology cannot properly utilize synthetic (mirror-image) fractions in the way natural complexes find their way into the biological reactions that are essential to tissue repair and the sustenance of life. A synthetic vitamin fraction can only be utilized for a drug or pharmacologic effect. Drugs are poisons, by definition. All pharmacologic therapies induce side effects and high-dose nutritional therapies are no exception. Only synthetic supplements achieve drug status. The effect of a drug is palliative not curative. Palliation means a masking or covering-over of symptoms, while the disease process remains unchanged or progressively gets worse for lack of attention. "It is better to have no vitamins at all than a mixture of crystalline-pure vitamin fractions, with one or more members of the complex missing."

Natural complexes differ from synthetic or crystalline vitamins in many ways:

- ☆ Natural complexes are colloidal, protein in nature and usually exist as enzymes or coenzymes.

- ☆ The crystalline vitamin can't be separated from its protein part without destroying its biological activity.
- ☆ The natural "live" complex carries trace mineral activators...without which the enzyme fails as a biochemical catalyst.
- ☆ Crystalline vitamins, taken into the body, must recombine (if possible) with their natural required components (robbed from the body's reserve) before they can function as a vitamin. Most, if not all of the crystalline component is lost through the kidneys.

Vitamins are not meant to be eliminated in urine any more than silverware is to be thrown out with the leftovers. Often you hear, "But, I feel better when I take them." The reason is that synthetic vitamins act as poisons in the body. Since most poisons stimulate circulation in the human body for the purpose of moving blood through the kidneys, more often to filter out the said toxin as quickly a possible... the speeded-up circulation also delivers more blood and more oxygen to the brain with a created sense of euphoria. Natural and low-potency vitamin complexes do not and will not produce euphoria, any more than eating a (non-poison-sprayed) carrot.

A synthetically-derived substance may cause a reaction in a chemically susceptible person when the same material of natural origin is tolerated, despite the two substances having identical chemical structures. The layman who doesn't understand the biochemical nature of true nutritional changes, is easily deceived by the pseudo-nutritional therapist and believes the initial response (drug effect), to the massive dose of one or more prescribed megavitamins, represents that what is being taken is exactly what is needed. Nutritional therapists know after a time on synthetic vitamin fractions, the biochemical reaction is reversed, the euphoria ends and the original symptoms return; except amplified by vitamin poisoning. Also, the diagnosis and evaluation of symptoms are now clouded by the weeks, months or years of improper supplementation. Why the great mystery about all of this?

Physicians are not required to understand nutrition to be licensed to practice medicine. It's not hard to understand why, when medical experiments are performed with synthetic vitamin fractions, these counterfeits fail to produce impressive results. But a reductionist seeks to understand the whole by dismantling its parts. This precludes the wholistic principle of synergism, that the whole is greater than the sum of its parts. For example, what if we tried to understand the properties of water (H2O) by testing hydrogen separately and oxygen separately. Presumably, we would learn enough about each to understand what their combined characteristics would be. Combined as 2 parts hydrogen and 1 part oxygen, H2O will extinguish fire. Separately, the elements are among the most flammable and explosive elements in the universe. The functions are exactly the reverse in their isolated state from their organically combined state. Similarly, vitamins function as biological mechanisms only when whole and complete, combined with their synergists, as in whole food. Isolated into synthetic chemicals, they fail as catalysts.

11.3.1 Categories of Supplements

The following is a summarization of the basic differences between the four categories of supplements available today.

11.3.2 Pure, Hypoallergenic or Natural

This is the most prevalent category of professional supplements available today. Contrary to popular belief, the vitamins and minerals used to create supplements in this category are not derived from consumable food or botanical sources. Instead, they are synthetically produced, fractionated or isolated from refined raw materials such as crushed rock, petroleum by-products and organic solvents to produce a pure, crystalline vitamin or mineral analogue. Although the final compound is an analogous structure to a specific vitamin or mineral, it is no longer attached or associated with any of the synergistic co-factors that the nutrient is inherently found within foods. As a result, pure vitamins and minerals have significantly reduced biological activity and a lower rate of utilization in the body.

11.3.3 Food-Based

This category of supplements is created from a physical mix of pure, fractionated nutrients with varying amounts of food powders or extracts. There are several different factors that need to be considered when determining the absolute nutritional value of a food-based formula. The primary factors that affect the quality of a food-based supplement are the amount, variance and form of the foods included in a particular formula. A premium food-base formula is one that provides a broad range of unadulterated whole food or plant concentrates (preferred over powders) in at least a 3:1 ratio of food to nutrients in a formula will significantly improve the bioavailability and biological activity of the vitamins and minerals. Delivering isolated vitamins and minerals in a premium food base is an energy-saving process, as the body will not waste vital energy stores to transform the vitamins and minerals into absorbable nutrients. Ascorbic acid is a good example of a chemically synthesized vitamin that has been vastly studied; it has demonstrated enhanced utilization when the nutrient is delivered with its natural co-factors, such as bioflavonoids, rutin and tyrosinase, which always co-exist with vitamin C in food. The utilization factor can be increased by as much as 3 to 5 times (from 10% for vitamins in the first category, to 3 - 50% for food-based).

To determine if a supplement has a food-base, review the label for chemical names, as well as a combination of food extracts or powders. To determine the ratio of nutrients to foods, add up the total milligrams of the listed nutrients and for the food extracts.

11.3.4 Biocultured Supplements

Biocultured supplements are a physical mix of food, pure isolated nutrients and may have one or more strains of yeast and/or probiotics. Once combined, this mixture is allowed to ferment in a controlled environment. The process of fermentation, also known as culturing, biologically alters the biochemistry and nutrient content of the food. Cultured foods have been shown to have numerous health benefits and have played a role in the diet of many cultures for centuries; however, the body cannot sustain optimal health on cultured foods alone. This is due to the loss of substances, nutrients or compounds that are protective of health, such as vitamin c. Depending on the strains used during the culturing process, the generation or inactivation of toxic substances may also occur. In addition, many proteins and enzymes are denatured during the culturing of biological activity of the nutrients.

11.4 WHOLE-FOOD SUPPLEMENTS

Many companies tout that their supplements are made from whole-foods; however, there are only a few companies that actually can substantiate this claim. Authentic whole food supplements are made with vitamins and minerals delivered in whole-food concentrates. The food concentrates selected have been grown in a controlled environment where budding plants are fed a nourishing broth enriched with bioactive peptide carriers. The peptide carriers act as chaperones, which govern a vitamin's or mineral's capacity to store the nutrient within its cells is successfully achieved, the growing process is complete, and harvesting is commenced using proteolytic enzymes to break down the cells' walls, and release the intra-cellular material.

The whole-food extract or concentrate produced is rich in a particular vitamin or mineral and is delivered in a synergistic state, with all of the plant's inherent food attachments intact and unadulterated. These attachments are vital co-factors facilitating the body's natural ability to recognize and utilize the nutrients present within the concentrate. Whole-food nutrients are recognized by the body as food, and therefore are better utilized and retained by the body when compared to any other supplemental form. Food is the complex form in which all nutrients are found in nature.

Supplements in this category will list the nutrient names--vitamin C, folate, calcium, etc.--and the mg. dosage of each, butthey will not denote a fractionated chemical name. Nutrients as they exist in food cannot be listed as one identifiable organic compound in this way, for the reason that they always exist within an interconnected weave of parts, and never as one isolated element. Nutrients in a whole-food state will be listed with their food source in the supplement facts panel such as vitamin C (Citrus sinensis)

11.4.1 Bioavailability Spectrum

On the low end of the spectrum you would have an inexpensive, store bought, hard-packed tablet. Something brightly colored that you can't break with your fingers. Such supplements have many fillers and binders, including ingredients that can be found in wood putty and tar. You can't take these on an empty stomach, or else you'll get an upset stomach. These are difficult to digest and breakdown, and often times your body doesn't at all. And these have been known to have been passed whole! For that reason you can understand why they have a bioavailability of 0-25%. So, just as a reminder, when you look at the daily values of nutrients on the label of these products, your body is not getting those percentages of nutrition, but rather up to 25% tops.

Moving up the bioavailability spectrum we come to the capsule delivery system. They have no fillers or binders, and you usually have to take larger quantities of these because they are not very concentrated. The bioavailability is a little better, but still only 40-60%. Getting higher up the bioavailability spectrum brings us to liquid supplements. These are better then the previous two in bioavailability and range between 80-90%, but many liquid herbs and oils do not taste very good.

11.4.2 Isotonic-Capable Supplements

Isotonic refers to "Iso" means "the same", and "tonic" means "pressure", so a solution being the same pressure as the fluids of the body. For example, an IV is isotonic. It is the proper ratio of nutrients to water so that it can be seamlessly integrated into the blood stream. An isotonic capable supplement works similarly. It is the proper measurement of nutrients to water to create an isotonic capable solution. When it is taken on an empty stomach, it is recognized as the same pressure as the stomach. The pyloric valve is signaled to open, and the full concentration of nutrients is delivered to the small intestine (our absorption site) where it is up to 95% absorbable within 5-15 minutes of consumption.

It is important to note that not all liquids are isotonic capable, again it is a specific ratio of medicine or nutrients to water which makes it isotonic. Everything that you ingest and put into your stomach has to be digested down to an isotonic state before it is released to the small intestine. When you eat some food and take a standard pill or capsule vitamin, it all remains in your stomach digesting for anywhere from 40 minutes to 4 hours. At that point a certain amount of dilution of nutrients has taken place. Some low quality pills do not even break down due to harsh fillers and binders, but still pass to the small intestine where none of the content of the supplement can be absorbed. This is why the bioavailability of pills and capsules are so significantly lower than isotonic capable supplements.

11.5 VITAMIN-A

Vitamin-A was the first fat-soluble vitamin to be discovered. In 1919, it was recognized that egg yolk contained a fat-soluble substance essential for life. In 1924, it was demonstrated that the eye disease, xerophthalmia, in children could be prevented by feeding them butter or cod liver oil. Vitamin-A was purified--separated from its natural constituents--and its refined chemical structure proposed in 1931. Synthesis, an artificial chemically-pure duplication, was accomplished in 1947 at Hoffman-LaRoche, a pharmaceutical company in Switzerland. Vitamin-A is actually a whole family of compounds, including retinol, retinal, retinoids, carotenes, and carotenoids, and always occurs in nature with synergists such as fatty acids, chlorophyll, other vitamins, enzymes, minerals, and trace elements. Retinol, retinal, and retinoids are found in foods of animal origin such as eggs, liver, fish, butter, and cheese.

The scientific term for vitamin-A is retinol, because of its presence in the retina of the eye. The role of retinol in vision was elucidated by a number of brilliant scientists, beginning in 1877 with a German, W. Kuhne, who discovered that the purple retinas from dark-adapted frogs turned yellow when exposed to light. The purple color is restored in a complex biochemical cycle involving vitamin-A, which makes vision possible. Other scientists demonstrated the role of vitamin A in cell differentiation, bone development, reproduction and immune system function. Dr. Weston Price confirmed the value of vitamin-A in traditional diets during his studies of primitive peoples carried out during the 1930s and 1940s. Due to the outstanding scientific work of these and many other researchers, the administration of cod liver oil to growing children--a tradition found among Arctic peoples such as the Scandinavians and Eskimos--became standard practice until after the Second World War.

In recent decades, much vitamin-A research has focused on its role in preventing cancer, and its use in combination with nontoxic therapies in the treatment of cancer. Dr. Max Gerson treated many cases of terminal cancer with excellent results, using raw liver juice, a rich source of vitamin-A. Modern confinement farming practices effectively prevent vitamin-A from incorporation into animal foods and the processing industry would rather use vegetable oils than animal fats. Some vegetable oils contain carotenes but they do not contain true vitamin-A. Only animal fats contain vitamin-A and vitamin-A is present in large amounts only when the animals have a source of carotenes or vitamin-A in the diet, such as green pasture, insects, and fish meal. Many popular books on nutrition insist that humans can obtain vitamin-A from fruits and vegetables. Even worse, FDA regulations allow food processors to label carotenes as vitamin-A. Vitamin-A from butter may be two or three times as potent, unit for unit, as vitamin-A from cod liver oil in some people.

The food industry, and the low-fat school of nutrition that the industry has spawned, benefit greatly from the fact that the public has only vague notions about vitamin-A. In fact, most of the foods that provide large amounts of vitamin-A--butter, egg yolks, liver, organ meats and shellfish--have been demonized. Under optimal conditions, humans can indeed convert carotenes to vitamin-A. This occurs in the upper intestinal tract by the action of bile salts and fat-splitting enzymes. Of the entire family of carotenes, beta-carotene is most easily converted to vitamin-A. The accepted ratio of how many units of beta-carotene is needed to produce one unit of vitamin-A is 6:1 and some research suggests an even higher ratio. This means that you have to eat an awful lot of vegetables and fruits to obtain even the daily minimal conversion. But the transformation of carotene to retinol is rarely optimal. Strenuous physical exercise, periods of physical growth, pregnancy, lactation and infection are stresses that quickly deplete vitamin-A stores.

Diabetics and those with poor thyroid function, which could include at least half the adult U.S. population, cannot make the conversion. In fact, the thyroid gland requires more vitamin-A than the other glands, and cannot function without it (mercury toxicity diminishes the thyroid and pituitary glands' functions). And a diet rich in vitamin-A will help protect the diabetic from the degenerative conditions associated with the disease, such as problems with the retina and with healing (mercury binds with the sulfur in insulin). Children make the conversion very poorly and infants not at all--they must obtain their precious stores of vitamin-A from animal fats--yet the low-fat diet is often recommended for children. Vitamin-A from animal sources (retinol) is primarily absorbed within three to five hours after ingestion in the upper intestinal tract; it is here that the fat-splitting enzymes and bile salts convert carotenes into usable nutrients, though conversion and assimilation takes up to six or seven hours. Carotenes are converted by the action of bile salts (mercury inhibits production of taurocholic acid, a bile salt) and very little bile reaches the intestine when a meal is low in fat. The conversion of carotenes is stimulated by thyroxine, a hormone secreted by the thyroid gland. Zinc, iron, and vitamin-E status also influences carotene conversion. Once converted, the carotenes are absorbed in the same way as retinol and allocated by the liver. Unabsorbed carotene is excreted in the feces.

Butterfat stimulates the secretion of bile needed to convert carotenes from vegetables into vitamin-A, and at the same time supplies very easily absorbed true vitamin-A. Polyunsaturated oils also stimulate the secretion of bile salts but can cause rapid destruction of carotene unless antioxidants are present. It is very unwise, therefore, to depend on plant sources for vitamin-A. This vital nutrient is needed for the growth and repair of body tissues; it helps protect mucous membranes of the mouth, nose, throat and lungs; it prompts the secretion of gastric juices necessary for proper digestion of protein; it helps to build strong bones and teeth and rich blood; it is essential for good eyesight; it aids in the production of RNA; and contributes to the health of the immune system. Strenuous physical exercise, excessive consumption of alcohol, excessive consumption of iron (especially from "fortified" white flour and breakfast cereal), use of a number of popular drugs, excessive consumption of polyunsaturated fatty acids, zinc deficiency and even cold weather can hinder the conversion of carotenes to vitamin-A, as does the low-fat diet.

Natural vitamin-A provided by liver, eggs, butter, cream and cod liver oil is well recognized as providing excellent protection against birth defects. Vitamin-A deficiency in pregnant mothers results in offspring with eye defects, hydrocephalus, displaced kidneys, harelip, cleft palate and major malformations of the heart and large blood vessels. Vitamin-A stores are rapidly depleted during exercise, fever and periods of stress. Vitamin-A deficiencies are widespread and contribute to high infant mortality, blindness, stunting, bone deformities and susceptibility to infection. These occur even in communities that have access to plentiful carotenes in vegetables and fruits. It is required for cellular differentiation (determines the function that cells will have): this assures that the cells which are lost from natural turnover, stress, insult, injury, disease, etc. are reproduced in the exact same form as the ones being replaced. Growing children actually benefit from a diet that contains considerably more calories as fat than as protein. Generous amounts of vitamin-A insure healthy reproduction and offspring with attractive wide faces, straight teeth and strong sturdy bodies. This vitamin assists in normal pregnancy, embryonic development, successful reproduction, fertility, lactation (nursing), and reproductive organ function, spermatogenesis, to adrenal, thyroid, and other gland functions. Vitamin-A is important for normal eyesight; necessary for night vision. A high-fat diet that is rich in vitamin-A will result in steady, even growth, a sturdy physique and high immunity to illness. Vitamin-A-rich foods like liver, egg yolk, cream and shellfish confer resistance to infectious diseases in children and prevent cancer in adults.

Children with measles rapidly use up vitamin-A, which can result in irreversible blindness. An interval of three years between pregnancies, allows mothers to rebuild vitamin-A stores so that subsequent children will not suffer diminished vitality. Kwashiorkor is as much a disease of vitamin-A deficiency, leading to impaired protein absorption, as it is a result of absence of protein in the diet. High-protein, low-fat diets are especially dangerous because protein consumption rapidly depletes vitamin-A stores. Children brought up on high-protein, low-fat diets often experience rapid growth. The results--tall, myopic, lanky individuals with crowded teeth, and poor bone structure, are commonplace in America. High-protein, low-fat diets can even cause blindness. Scarcity of good quality raw dairy products, a rejection of organ meats as old fashioned or unhealthful, and a substitution of vegetable oil for animal fat in cooking all contribute to the physical degeneration and suffering of Third World peoples. Supplies of vitamin-A are so vital to the human organism that mankind is able to store large quantities of it in the liver and other organs. Even people who can efficiently convert carotenes to vitamin-A cannot quickly and adequately replenish vitamin-A stores from plant foods. Many factors interfere with its absorption and utilization. Inadequate fat in the diet, poor production of bile salts, low enzyme status, and compromised liver function can all interfere with the uptake and usage of vitamin-A, especially when given as a supplement in the form of retinol, rather than as a component of whole foods. One tablespoon of cod liver oil contains at least 15,000 IU and one serving of liver (should be from organically raised animals) can contain up to 40,000 IU vitamin-A. Foods high in vitamin-A are especially important for diabetics and those suffering from thyroid conditions. Weston Price considered the fat-soluble vitamins, especially vitamin-A, to be the catalysts on which all other biological processes depend. Efficient mineral uptake and utilization of water-soluble vitamins require sufficient vitamin-A in the diet.

11.5.1 Synthetic Vitamin-A

Synthetic vitamin-A interferes with the proper utilization of natural vitamin-A from foods. Pure retinol is added to many fabricated foods like margarine, white flour, extruded breakfast cereals and pizza. A synthetic retinoic acid derivative,isotretinoin (Accutane) is used to treat acne as well as psoriasis, cancer, and oral leukoplakia. Children whose mothers took Accutane during pregnancy were born with nervous system defects, brain damage, congenital heart disease, and more. Depression, crying spells, malaise, and forgetfulness are reported human symptoms. Inflammation of the lips (cheilitis) develops in 90% of the users; conjunctivitis develops in 40%. Muscle damage has also been added to the list. The synovial membranes in joints are supported by vitamin-A. The vitamin-A complex is supportive to the adrenal glands which produce cortisone and other needed hormones; and vitamin-A plays an important role in any inflammation and repair process. Vitamin-A is important to the external and internal mucous membrane tissues as well as to the immune system. However, synthetic retinoic acid metabolites do not reveal any specific supportive mechanism and results are not seen with synthetic, altered fractions.

The whole vitamin-A complex is known to be very functional, but when distilled or synthetic fractions are used, roles and results are questionable at best. Megadoses of synthetic vitamin-A can result in hypertrophy (excessive growth) of the liver and spleen, with elevation of AST (aspartate aminotransferase), a liver enzyme. Liver fibrosis (abnormal formation of fibrous tissue), fat deposition, obstruction of portal blood flow with portal hypertension (high blood pressure) and sclerosis(hardening) of the blood vessels have been found at liver biopsies. Excess synthetic vitamin-A is also associated with calcium deposits in soft tissues, contributes to attacks of gouty arthritis, hair loss, irregular menses, emotional lability, insomnia, restlessness, night sweats, and more. Vitamin-A is essential to spermatogenesis, the production of sperm in men, and it requires alcohol dehydrogenase, an enzyme to convert retinol to retinal. This enzyme may require the trace element, zinc, as well. Excessive doses of synthetic fractions of vitamin-A can convert to a compound that can be more damaging than a deficiency in its effect on the germinal epithelium of the testes. Vitamin-A is present in almost twice the amount in seminal fluid of highly fertile men compared with that of men deficient in sperm. Zinc increases vitamin-A uptake by the seminiferous tubules of the testes. Vitamin-A is bound to a protein, called retinol-binding protein. The necessity of enzymes, trace mineral activators, amino acids, and other constituents stresses the point that the whole food complex is important and effective whereas an isolated portion cannot produce the same results. Excessive fractionated "vitamins" A and E may produce testicular hemorrhaging. Zinc is required to maintain normal concentrations of vitamin-A in the blood plasma.

Diabetic persons tend to have decreased levels of zinc. Since a zinc deficiency may lead to difficulties mobilizing vitamin-A from the liver for use elsewhere in the body, vitamin-A deficiency will worsen the condition. Another essential constituent is fat, as vitamin-A occurs in foods with fatty acids and is a fat-soluble vitamin. It is absorbed much better when natural, unaltered fats are also present. Other fat-soluble vitamins occur with and work with whole vitamin-A complex such as vitamin E complex and essential fatty acids. When vitamin E is deficient, one develops low blood levels of vitamin-A; when vitamin E is restored to the diet, vitamin-A levels become normal. Yet, using chemically-pure fractions of "vitamins" A and E in large doses creates more problems, more imbalances. Fractionated alpha-tocopherol (so-called vitamin-E) can supplant vitamin A in the body, reducing its effectiveness. Administration of the whole food complex brings excellent responses. Stubborn cases of gastrointestinal irritation and even ulceration in the digestive tract that failed to respond to other treatments have totally resolved after use of real vitamin-A complex. Persons with cystitis (inflammation of the bladder) experience relief after a relatively short time taking food concentrates of Vitamin-A and other supportive nutritional complexes. Edema caused by liver pathology is another instance where vitamin-A complex is of particular value; the liver is the chief location of vitamin-A stores. Colds, chronic rhinitis, sinusitis, psoriasis, and many other conditions--especially those involving epithelial tissues--are responsive to the whole vitamin-A complex.

A new strategy that has been the development of and promotion of genetically engineering rice to produce carotenes. Golden rice containing carotenes can't provide true vitamin-A to the world's children but it will further the trend of pushing their parents off the farm and into ghastly slums. The most effective response that consumers have is to boycott processed foods!

11.6 VITAMIN B-COMPLEX

The B-COMPLEX factors naturally divide themselves into two groups. One group consists of the factors that are alcohol and water soluble. This includes B1, B12, Pantothenic Acid and Adenine, but the primary factor in this group is called Vitamin B4, an anti-paralysis vitamin. It is one that has not yet been recognized by the FDA as being necessary in human nutrition. They say it helps pigeons, but not people. So vitamin B4 is a nerve-promoting factor, it promotes the transmission of nerve impulses. This would be needed with heart conditions involving irregular heartbeat, fibrillation, or heart blocks. The B4 factor also causes vessels to constrict. It's very useful for people with low blood pressure--the person whose blood vessels are just hanging open, with no tone. The B4 puts strength in the vessels and snaps them into action by improving the transmission of nerve impulses to the vessels. For this reason, vitamin B would be contraindicated in cases of organic heart trouble, such as angina. Here you would need the G factor. The G factors are not soluble in absolute alcohol. Complex G contains the lipotropic factors that metabolize fat.

Complex-G contains B2, B3, B6, Choline, Betaine, Inositol, Folic Acid, Biotin and PABA, and is a vasodilator--it opens up the blood vessels by relaxing the nerves in the vessels. So complex G is for a person with high blood pressurethe tense, coronary type person. It is important to have these factors separated and useable either by themselves or together. This benefit is not achieved with megadoses of synthetic so-called Vitamin-B. Today, the average practitioner of the medical arts, if sufficiently brain washed by the medical-drug-food-dairy-chemical combine of disinformation, cannot conceive that any person could suffer from B-complex deficiency. Even if they know that refining and processing of food, chemical fertilizers and food additives, all diminish vitamin content and nutrient quality. Because these "foods" are "enriched" with synthetic vitamins. University biochemists and leading health authorities parrot the phrase: "There is no difference between natural and synthetic vitamins." Below are diagrams of the structural formulae of NATURAL vitamin B1 (left) and synthetic B1 (right). Show any high-school chemistry student with a passing grade the diagrams and tell them they are the same, and that the body does not know the difference. Now, leading nutritionists from the university fountainheads of knowledge will tell you: “there is no difference.”

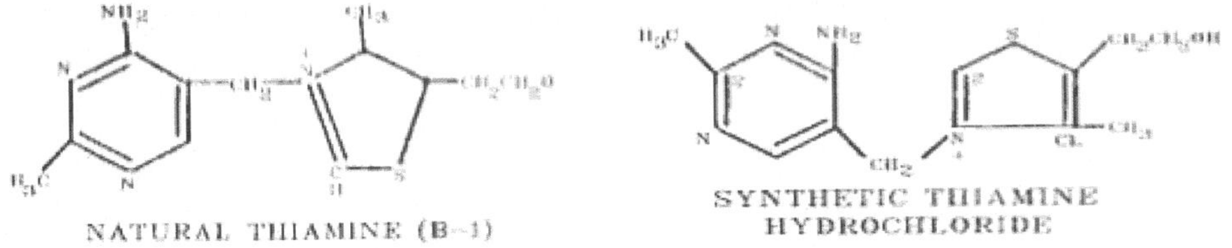

Fig sulfur-containing molecules

Thiamine is one of two vitamins in the body that contains a sulfur entity; the other is biotin. The high binding affinity that mercury has for the sulfur-containing molecules (thiols & disulfides) gives it the biochemical potential to affect the functions of this critical vitamin and affect critical energy processes in the body. Not only does thiamine have a sulfur atom in its structure to start with, but its involvement as a coenzyme in critical biochemical processes is connected through transfers involving sulfur molecules. Out of about two-dozen substances produced by the metabolism of thiamine in the body and excreted in the urine, science has only identified the function of six of them thus far. There is a lot going on metabolically with vitamin B1, that mercury can affect, that haven’t even been considered as potential problems because the complete biochemical function of the vitamin has not been identified.

Thiamine is involved in many enzyme reactions in the body. As a coenzyme, B1 is involved in the intermediary metabolism of carbohydrates in all cells of the body. For natural vitamin B1 to perform as a coenzyme, it must be phosphorylated or combined with two molecules of phosphoric acid. As such, it then functions as thiamine pyrophosphate (TPP). In the biochemically active form, it is also called diphosphothiamine (DPT). Better textbooks of biochemistry use the TPP designation as preferable. Early symptoms of thiamine deficiency, are intriguingly similar to mercury toxicity and are non-specific and hard to pinpoint: fatigue, anorexia, weight loss, gradual loss of muscle strength, peripheral neuritis (numness or increased sensitivity, tingling in the extremities, loss of reflexes), irritability, confusion, depression, lack of initiative, poor memory, gastrointestinal problems (including constipation), and low body temperature. Cardiovascular symptoms can include edema of the ankles, feet and legs, palpitation, tachycardia, abnormal electrocardiograms, decreased blood pressure, and difficulty in breathing after exertion. Thiamine participates in the metabolism of tryptophan to form nicotinic acid. Because of this, a deficiency of thiamine can cause a deficiency of vitamin B6, which is also required in the metabolic pathway of tryptophan. Thiamine is a cofactor in steroid hydroxylation, fatty acid synthesis, and glucose metabolism in the pentose phosphate cycle, which is another energy system. Mercury inhibits glucose-6-phosphatase, a key enzyme in the glucose cycle. The proportion of glucose metabolized by this route is high in the lactating mammary gland, the adrenal cortex, leukocytes and erythrocytes. Each one of the mentioned also are affected by mercury in ways other than glucose metabolism. Not only mercury, but arsenic and lead impact on the biochemical pathways of these vitamins and minerals. If a person is deficient in thiamine or B1, most likely they are also lacking in all members of the B-complex.

11.6.1 Symptoms

- ☆ Weakness and fatigue, insomnia, abnormal fears
- ☆ Indigestion (hypochlorhydria), dizziness, rage
- ☆ Decreased appetite, nervousness, and hostility
- ☆ Craving for sweets, instability, depression
- ☆ Neuralgia to neuritis, forgetfulness, and anxiety
- ☆ Muscular soreness, vague fears, apprehension
- ☆ Headache, uneasiness

Vitamin B deficiency leads to heart block and a remarkably slow pulse... Vitamin B deficiency is known to produce a loss of appetite through its effect of causing a loss of tone of the stomach muscles. This loss of tone also may affect the heart and muscular walls of the arterioles--resulting in heart weakness and edema, respectively. Thus, under stress, exaggerated mental activity requires additional glucose for the brain cell to function. Not only must the brain cell have extra glucose to cope with stress, the cytoplasm and mitochondria of brain cells (neurons), must have all of the members of the B-complex in their biologically-active-form the coenzymes required for anaerobic and aerobic glycolysis.

Minerals and trace elements must also be available optimally. Phosphorus, Potassium, Magnesium, Manganese and Calcium play important roles in the conversion of sugar into energy, within all cells of the body. Being deficient in the B-complex factors, one cannot properly convert sugar or glucose to pyruvic acid for the anaerobic glycolytic pathway of sugar metabolism in the cell cytoplasm. Without B1, pyruvic acid cannot be converted to coenzyme A and enter the mitochondria to participate in the citric acid cycle and form ATP. As a result, pyruvic acid accumulates in nervous tissue and piles up on nerve endings causing irritation. Then the cells signal for help.

The adrenal medula comes to the rescue, increases its function and hormone output...and, as a result, provides the mechanism and the stimulus for producing muscle and brain energy by gluconeogenesis. As long as adrenal hormone output is sufficient, one can survive all the symptoms of B-complex deficiency. The adrenal glands need protein, minerals and vitamin C (not ascorbic acid), especially tyrosinase (organic copper), to produce catecholamines. Without that extra energy, the person will flee, or cry, or hide. Emotionally they usually hide or separate from reality... the classical nervous breakdown, or nervous exhaustion. When this inevitable point is near at hand, the person not only needs total nutritional support, but also adrenal support. Besides those nutrients mentioned in the glycolytic phase, the Krebs energy cycle uses B3 five times, B1 two times, Pantothenic acid two times and Biotin three times. Biotin, normally synthesized in the intestines, is needed to transport CO2 from the cells into the circulation. Heavy metal toxicity blocks Biotin function, while antibiotics and chlorine in water, stop or restrict Biotin production in the intestinal tract. In the anaerobic phase of glycolysis, Fluoride blocks the very important enolase reaction, which impairs glucose metabolism. B2, Riboflavin or "G" combines with several enzymes and enzyme systems forming many flavo-proteins. It is present in retinal pigment contributing to visual acuity. Deficiency signs and symptoms include:

1 Corneal vascularization (red eye)	8 Myelin degeneration
2 Corneal cloudiness and ulceration	9 Incoordination
3 Cataracts	10 Loss of strength in arms and legs
4 Photophobia (burning and itching of eyes)	11 Central neuritis
5 Abnormal pigmentation of iris	12 Fissures at corners of the mouth
6 Stomatitis with lesions	13 Scaliness and greasiness at nose & ears
7 Glossitis	14 Impaired red blood cell formation

B^3, or Niacinamide. The acid niacin or nicotinic acid as found in plants is a vasodilator...the amide form as found in animals is not. B^3 deficiency signs

1 Erythematous cutaneous lesions aggravated by sunlight	10 Failing vision
2 Anorexia, nausea and vomiting	11 Excessive salivation
3 Dilation of blood vessels (aggravated by Niacin)	12 Burning hands and feet
4 Epithelial atrophy of tongue	13 Pain in calves
5 Headaches, dizziness, insomnia, depression, impaired memory	14 Numbness and weakness of extremities
6 Enteritis to ulcers	15 Difficulty in walking
7 Fatty infiltration of liver	16 Absent knee jerk

8 Watery to bloody stools	17 Psychoneurosis to stupor or mania
9 Myelin degeneration (motor and sensory)	

B6, or pyridoxine as a pyridoxal-5-phosphate, functions as a coenzyme in the metabolism of amino acids, fats and fatty acids. Transports amino acids and potassium into cells. Deficiency signs and symptoms include:

1 Seborrheic dermatitis around eyes, eyebrows and angles of mouth	5 Lethargy and mental confusion
2 Glossitis and conjunctivitis	6 Decreased lymphocytes
3 Severe sensory neuritis (a.) Numbness and tingling in hands & feet (b.) Hyperesthesia	7 Increased Eosinophils
4 Anorexia, nausea, and vomiting	8 Elevated B.U.N. (Blood Urea Nitrogen)

11.7 FOLIC ACID

Folic acid deficiency is one fo the most common vitamin deficiencies. Many of the symptoms are similar to those of B12deficiency. As much as 100% may be lost if foods are improperly stored, cooking water is discarded, or foods are reheated or overcooked. Many medications, including aspirin and anticonvulsants, may also interfere with folacin absorption and metabolism, further decreasing its availability. Folic acid is involved in the metabolism of pantothentic acid and is required for its utilization. It is through this participation that it is involved in antibody production of the immune system. Folic acid also has a relationship with other vitamins including vitamin C and E. In the body, folic acid is rapidly converted to the biologically active form tetrahydrofolic acid (THFA) in the presence of NADPH (niacin's coenzyme form) and ascorbic acid. In conjunction with vitamin B12, THFA participates in amino acid conversions and the methylation of choline, methionine, serine, and histidine. Folates carry out their metabolic function as a carrier of one-carbon units in the tetra-hydro form. It is this capability to accept one-carbon units, which synthesize various coenzymes, that makes folates so important and involves them in so many metabolic functions. The liver actively reduces and methylates folates which are then transported into the bile for reabsorption by the gut and subsequent delivery to the tissues. This important pathway provides as much as 200 micrograms or more of folate each day for recirculation to tissues. The folates are a family of coenzymes that function in association with their respective enzymes to accomplish many intracellular metabolic functions. Methyl-THFA is required as the methyl donor in the conversion of homocysteine to methionine (methyl methionine).

This reaction utilizes vitamin B12 as a cofactor and also requires vitamin B6. Folates are also involved in the conversion of the amino acid serine to glycine, and histidine to glutamic acid. It is also necessary in the synthesis of glucose, and DNA. Folic acid and vitamin B12 are both involved in synthesizing DNA. In megaloblastic anemia (folate deficiency), certain steps in this vital process of DNA synthesis are reduced, causing an arresting of the synthesis of red blood cell replicaiton. The form of folacin most commonly found in the liver and serum is methyl-THFA. Methyl-THFA can only recycle through a B12-dependent pathway. If a B12 deficiency exists, folic acid is trapped as methyl-THFA and is useless to the body. A deficiency of either vitamin B12 or folic acid will result in identical hematological symptoms.

A vitamin B12 deficiency impairs the conversion of homocysteine to methionine. The enzyme methionine synthetase requires vitamin B12. Therefore, failure of methionine synthesis leads to inadequate formulation of THFA. Methionine deficiency has been equated with B12deficiency. The phagocytic function of polymorphonuclear leucocytes (PMNs), and to a lesser degree bactericidal function, are adversely affected by folic acid deprivation. Folic acid requirements increase during pregnancy, and low tissue-folate levels are a predisposing factor in the bacteriuria (bacteria in the urine) of pregnancy. There is great potential for mercury and lead to be involved in folate deficiency. The involvement of chronic inhalation of mercury vapor from mercury amalgam dental fillings can be an etiological factor in thyroid dysfunction involving folic acid, B12, and thyroxine production.

11.8 PANTOTHENIC ACID

The physiological active form of pantothenic acid is coenzyme A, which is required for many different enzymatic actions. It is at the center of energy metabolism and fat, acetylcholine, and antibody synthesis. Coenzyme A, which is synthesized from pantothenic acid, is abbreviated as CoA or CoA-SH, and functions as a transient carrier and a cofactor for a variety of enzyme-catalyzed reactions involving the transfer of acetyl (two carbon) groups. The CoA-SH designation reflects the fact that the coenzyme A molecule has a reactive thiol (-SH) group, to which acyl-groups become covalently linked to formthioesters during acyl-group transfer reactions.

All known acyl derivatives of CoA and related pantetheine derivatives are thiol esters. Because of these derivatives, CoA is involved in the metabolism of carbohydrates, lipids, protein, and porphyrin. It is also involved in the synthesis of fatty acids, cholesterol, citrate, acetoacetate, and the neurotransmitter acetylcholine. Pantothenic acid is an essential component of the brain and must enter the brain and cerebrospinal fluid from the blood. In the brain, as part of CoA, pantothenic acid is involved in many important reactions. Symptoms identified with pantothenic acid deficiency such as fatigue, headache, insomnia, nausea, abdominal cramps, occasional vomiting, paresthesias of the hands and feet (burning foot syndrome), muscle cramps, and impaired coordination.

Pantothenic acid is required for the synthesis of cholesterol precursors for sterol hormones in the adrenal cortex, and a pantothenic deficiency can produce cortical necrosis. Conversely, pantothenic acid stimulates the adrenal glands and increases production of adrenal hormones. Mercury can inhibit production of adrenal hormones and cause adrenal over-growth. Mercury causes a direct defect in adrenal steroid biosynthesis with physiological consequences being lowered plasma levels of corticosterone and elevated concentrations of progesterone anddehydroepiandrosterone, all abnormal steroid hormone profiles. Mercury's affinity for thiols, sulfhydryl groups and the sulfur molecule, wherever they are biochemically available in the body, gives it the potential to affect countless metabolic functions. Symptoms produced by a pantothenic deficiency are all also documented as effects produced by mercury toxicity.

11.9 VITAMIN B6 (PYRIDOXINE)

Vitamin B6 participates in the metabolism of protein, carbohydrates, and lipids, and is a coenzyme constituent in the formation of erythrocytes and in amino acid metabolism. The conversion of methionine to cysteine is dependent on vitamin B6. The three forms of vitamin B6 that occur in foods are pyridoxine, pyridoxal phosphate, and pyridoxamine phosphate. The physiologically active forms of vitamin B6 are pyridoxal phosphate and pyridoxamine phosphate. All three forms are converted in the body to pyridoxal phosphate. Pyridoxal phosphate is involved with a number of enzymes in catalyzing reactions of amino acids. The most common of these are called transaminations, in which an amino group is transferred to a carbon atom. In these reactions, pyridoxal phosphate serves as a transient carrier of the amino group from its donor to the amino group acceptor. The derivative of this intermediate action is pyridoxamine phosphate which then donates its amino group to the carbon atom of alpha-keto acid, after which it reverts to its pyridoxal phosphate form. In the biochemical reactions related to the conversion of methionine to cysteine, there is a need to transfer a sulfur component, and an amino component, which would make the process doubly vulnerable to mercury because its highest binding affinity is for the sulfhydryl group and then the amino group. Further, mercury can bind to phosphate groups, which gives it a total of three possible mechanisms to affect the metabolism of pyridoxal phosphate, the biologically active form of vitamin B6.

One of the functions of vitamin B6 is called transsulfuration. This is the process whereby one sulfur molecule can replace another and represents an important biochemical function for utilization of sulfur. When pyridoxine is deficient in the body, methionine metabolism is altered. The pancreas is very active in transsulfuration and is known to have a rapid uptake of methionine, because of this, mercury can be involved in pancreatic dysfunction. Cystathionine is a major constituent of the pool of free amino acids in the brain. This fact raises the question of neurochemical involvement. Although vitamin B6 is not synthesized in the brain, it readily enters the cerebral spinal fluid and brain from the plasma.

Once within the CSF, B6 can enter brain cells. The holoenzyme needed to convert cystathionine to cysteine contains B6 in its pyridoxal phosphate form. It is relatively easy to deplete brain B6. Any substance that impedes or inhibits the metabolic functions of vitamin B6 in humans could have some very serious consequences. Symptoms of vitamin B6 deficiency, like those of micromercurialism, can be nonspecific and hard to pinpoint. They include weakness, mental confusion, irritability and nervousness, insomnia, poor coordination in walking, hyperactivity, convulsions, abnormal electroencephalogram, declining blood lymphocytes and white blood cells, anemia, and skin lesions. Mercury has the biochemical capability of reducing the availability of, or inhibiting the function of vitamin B6. There is a positive correlation between brain mercury levels and the numbers and surfaces of amalgam fillings.

11.10 VITAMIN B12

Vitamin B12 has the most complex structure of all vitamins and is also unique in that it is the only vitamin containing a metal ion. Cobalamin is the generic name of the vitamin because of the presence of the metal ion cobalt in its structure. Vitamin B12 was isolated in 1948. Merck and Company is the holder of various patents of procedures for producing synthetic B12and owns the product claim. B12 functions primarily as an essential coenzyme for the normal metabolism of all cells, including those of the gastrointestinal tract, bone marrow, and nervous tissue. It is also a coenzyme in the synthesis of red blood cells and in the maintenance of nerve cells, and is involved with protein, lipid and nucleic acid synthesis. It is considered necessary for growth. The coenzyme of vitamin B12 is a carrier of methyl groups and hydrogen and is necessary for protein, fat, and carbohydrate metabolism. Coenzyme B12, methylcobalamin, functions as a methyl-group donor to form methionine from homocysteine. Vitamin B12 coenzyme deoxyadenosylcobalamin functions in the conversion of methylmalonic acid to succinic acid. Methionine and B12 coordinate in the regulation of folate metabolism. A deficiency of either vitamin results in defective synthesis of DNA in any cell that attempts chromosomal replication and division.

Chronic mercury inhalation from mercury amalgam dental fillings, with its great affinity to bind to methionine and cysteine has the potential to decrease the availability of these amino acids and affect the metabolism of both vitamin B12 and folate in man. Methionine is needed in choline synthesis, which means that vitamin B12 plays a secondary role in this lipid pathway. A choline deficiency that causes fatty liver can be prevented by vitamin B12 or the other methyl donors--betaine, methionine, and folic acid. Impaired fatty acid synthesis, observed in vitamin B12 deficiency conditions, can result in impairment of brain and nerve tissue. The insulation around nerve cells, the myelin sheath, is misformed in vitamin B12deficiency, contributing to faulty nerve transmission.

Neurological disturbances result from prolonged vitamin B12 deficiency. Inorganic mercury inhibits outgrowth of nerve fibers and the development of glial cells, and also depresses the outgrowth of fibroblasts. Methyl-B12 is able to inhibit the neurotoxic effects of organic mercury but not those caused by inorganic mercury. Mercury vapor released from dental amalgam fillings is inorganic mercury. Very low concentrations of mercury and arsenic are able to inhibit nerve outgrowth, and lead directly interferes with the process of myelination. Proper DNA replication is dependent on the function of coenzyme vitamin B12 as a methyl-group carrier. Methyl-B12 may play an important role in immune regulation. Methyl-B12 is able to enhance the activity of helper T-cells for immunoglobulin synthesis of B cells. The presence of methyl-B12 significantly potentiates the induction of suppressor cells. Polymorphonuclear leukocytes (PMNs) are white blood cells that function in a phagocytic capacity (capable of destroying foreign substances). Low levels of mercury inhibit the metabolic reaction called respiratory burst in human PMNs. Mercury acts as an interference to normal biological functions involving the formation and/or function of sulfur containing proteins in cells, enzymes, and hormones. Biological damage will occur long before the appearance of any clinically observable signs and symptoms of mercurialism. Known causes of B12malabsorption are a lack of intrinsic factor and hydrochloric acid in the stomach, the removal or disease of the second portion of the ileum, and competition for B12 by microorganisms or parasites.

There is far more unknown than known about B12 and its metabolism in the body. There are several vegetarian food sources of B12. While these sources may not be constant, it appears that they occur with sufficient frequency to supply the minute quantities of B12 that are needed by those who are not abusing their health. These include: wheat, soybeans, various common greens, olives, fruits, and many other foods that occasionally have B12 either in or on the food. Vitamin B12has been found in roots and stems of tomatoes, cabbage, celery, kale, broccoli and leeks. It has been found in the leaves of kohlrabi. Seaweed and alfalfa contain B12 also. It is difficult to avoid vitamin B12. A further source of B12 is bacterial growth in the mouth (around the teeth and gums), in the nasopharynx, around the tonsils and tonsillar crypts, in the folds at the base of the tongue, and in the upper bronchial tree. Up to 0.5 micorograms daily can be obtained from this source. It's likely that this source alone will supply sufficient quantities of B12 for the small requirement that a pure vegetarian has, especially considering his low protein intake, which further reduces the need for B12. It has also been shown that some bacteria, which may colonize the small intestine of man, can synthesize considerable amounts of biologically active forms of the vitamin. The terminal ileum is principally involved in absorbing B12. Since bacteria can form B12 above the terminal ileum, and since intrinsic factor, which is needed for most rapid absorption, is found in the ileum, there can be adequate absorption of vitamin B12 produced from these bacteria in the ileum in healthy people.

Individuals taking conventional diets need only about 0.1 micrograms of B12 per day even though conventional diets contain excess fat, animal protein and refined foods all of which increase the need for B12. True vegetarians require only 0.05 micrograms of B12 daily. An ounce of the roots of leeks, beets and other vegetables will provide 0.1 to 0.3 mcg. of B1, more than a day's supply. The routine use of vitamin B12 is not advised. Cases of increased cancer production in animals receiving high levels of B12 have been published. A group of French investigators reported a series of cases suggesting that B12 may stimulate multiplication of cancer cells and aggravate the disease. Patients with rheumatoid arthritis present serum B12 levels significantly higher than normal subjects. It is also known to be higher in patients with ulcerative colitis, leukemia and other serious illnesses. Breast-fed infants get only a tiny fraction of a microgram of B12 per day, even if their mothers are on a high intake. Yet, they can build neurological tissue, convert their hemoglobin from F to A type, and do other complex functions dependent on B12. Iron deficiency interferes with absorption of B12. Large doses of vitamin B1 can also destroy vitamin B12, as can the use of oral contraceptives. Lactose intolerance may increase B12 needs.

11.11 VITAMIN C COMPLEX

Another danger inherent to crystalline-pure, megavitamin usage is the consumption of ascorbic acid--the so-called vitamin C that is recklessly prescribed by the "pseudo-nutritional pharmacologists" (Physicians, Nurses, Nutrition consultants, registered dieticians, health food store employees, etc.), who have little, if any, realistic training in this most important field. Ascorbic acid is not vitamin C! Ascorbic acid is a fraction of the biologically utilizable Vitamin C complex. Most all ascorbic acid sold is synthetically produced from corn sugar and even though it is synthetically manufactured, it is called "natural" and "organic" because corn is found in nature.

Technically, anything with a carbon atom can be called "organic." The terms natural and organic are totally meaningless. This article alerts the general public to the fact that as little as 1,500 milligrams of ASCORBIC ACID, taken daily, depletes blood copper, leading to serious health problems. You jeopardize your important "fight or flight" (stress) response by depletion of copper. Just 500 mg. of ascorbic acid with 3 meals can do just that. The symptoms of clinical scurvy include swollen joints, muscular aches, bone pain, edema, weakness, fatigue, anemia, loose teeth, hyperkeratosis, impaired wound healing. Behavioral changes may include apathy, depression, and emotinal disturbances, weakening of the walls of blood vessels such as swollen and bleeding gums, ocular hemorrhages, bruising, and varicosities of small blood vessels which are seen under the tongue.

Subclinical vitamin C deficiency may exist in a large segment of the population, and can lead to impaired health and increased susceptibility to other diseases. Ascorbic acid therefore is incorrectly named. With isolated ascorbic acid (from Hungarian red pepper), he could not stop the capillary hemorrhaging, so characteristic of scurvy, that he could reliably influence with the C-complex. He went back to the laboratory and discovered vitamin P, the rutin factor of the C-complex, which exists more in buckwheat (grain and leaf) than in citrus.

The P-factor of the C-complex gives capillary and blood vessel wall strength, the J-factors, increase the oxygen-carrying capacity of the blood. The known physiological functions of vitamin C are synthesis of polysaccharides and collagen; formation of cartilage, dentin, bone, and teeth; antioxidant; absorption of iron; cold tolerance; maintenance of the adrenal cortex; metabolism of tryptophan, phenylalanine, and tyrosine; growth; wound healing; and maintenance of capillaries. Vitamin C is directly involved in proline and lysine hydroxylation; carnitine synthesis; dopamine hydroxylation; drug and cholesterol breakdown; sulphation; lymphocyte and neutrophil function; and folate reduction. Vitamin C has a protective effect against mercury poisoning. For many years, mercury diuretics were used extensively by the medical profession. The toxicity of these mercurial diuretics could be reduced if the patient was given vitamin C prior to or simultaneously with the mercurial diuretic. When vitamin C is metabolized, part of it is metabolized to vitamin C-sulphate, with the sulphate being derived from sulphur-containing amino acids such as cysteine. Vitamin C competes with certain drugs for sulphate conjugation, which could affect the pharmacological activity and toxicity of drugs. Mercury competes actively for the sulphur-containing amino acids cysteine and methionine, as well as the cysteine molecule of the tripeptide glutathione. Anyone with amalgam fillings should routinely eat a diet rich in sulphur-containing foods to offset the depletion of cysteine and methionine caused by the inhalation of mercury vapor being continually released from amalgam fillings. The problem of cysteine depletion is further aggravated by the presence of mercury, lead, arsenic, and cadmium, because they are all thiol sensitive.

Mercury is also known to inhibit collagen synthesis. The synthesis of collagen is impaired in vitamin C deficiency, due to lowered ability to hydroxylate lysine and proline. The reducing agent in the hydroxylation of both lysine and proline is the reduced form of vitamin C. Consequently, in vitamin C deficiency, the amount of effective collagen fiber present in connective tissue is reduced. In severely scorbutic patients, swelling, hemorrhages, and secondary bacterial infections of the gingival margins are common. It is not the actual deficiency of vitamin C that causes the inflammation, but the lack of real vitamin C, impairing the normal defensive responses of the mucous membranes. Thus, the massive gingival enlargement so characteristic of scurvy results from the combined effects of lack of vitamin C and nonspecific inflammation. The adrenal glands are one of the target glands of mercury deposition. They also contain the body's second highest tissue levels of vitamin C. The physical response to stress is an increased secretion of the hormones of the adrenal glands. Physical and mental stress increases adrenal activity which, in turn, depletes vitamin C from the gland. Humans replenish the adrenal stores of vitamin C by taking it from other stores in the body. If tissue values of vitamin C are low, there may be an insufficient amount available to replenish or satisfy the requirements of the adrenals. Thiol groups have a direct role in the preservation of tissue vitamin C and its reduced form, and any decrease in the biological reducing capacity of tissue thiols will result in lower levels of tisssue vitamin C. Under these conditions, normal adrenal hormone response may become inadequate. Prolonged exposures to low mercury concentrations depress the adrenal vitamin C content. The person with amalgam dental fillings who is being exposed to chronic intakes of mercury vapor is subjecting the adrenals to depletion by chemical stress. One function of the liver tissue is to reduce dehydroascorbic acid to ascorbic acid. This ability guarantees the preservation of the body's vitamin C reserves, so necessary to normal function.

The body doesn't want ascorbic acid per se; it wants the C-complex being preserved by ascorbic acid. Once in the body, the body sheds the ascorbic acid, the preservative wrapper, the way you shed the peel of an orange; takes the rest of the C-complex and places it as the lance of the lymphocyte; and kills bacteria with it. A lymphocyte unarmed with C complex will fail in its function, it will be impotent, it won't be able to destroy infectious organisms. Lymphocyte potency is dependent on copper, the core trace mineral of the tyrosinase enzyme. The reason one gets any relief from taking ascorbic acid is that ascorbic acid lowers the pH to the acid side of the pH scale. It acidifies the body, creating an unfavorable medium of growth for pathogenic bacteria. Acting in this way, ascorbic acid is not producing a "vitamin effect" but rather an "acidifying effect." Most infectious pathogenic bacteria thrive in an alkaline pH. The following bacteria, all well-known enemies of modern science's war on bacterial infection, grew optimally on alkaline media of pH 7.4 and above: staphylococcus (staph infection), streptococcus (strep throat), pneumococcus (pneumonia), h. influenza (the flu), meningococcus (meningitis), corymbacterium diptheria (diptheria), clostridium tetani (tetanus), and others.

Apple cider vinegar is 5% malic acid, a natural colorless crystalline acid normally found in the body. It can be taken with water during meals, and is much less expensive than ascorbic acid and does a better job in lower doses at acidifying the body, without causing rebound deficiencies as ascorbic acid does. Ascorbic acid has long been recognized as a contributing factor in tyrosine (copper) depletion. Tyrosinase is absolutely essential in the biosynthesis of catecholamines. The adrenal medulla supplies the catecholamines, epinephrine and norepinephrine which are biosynthesized from phenylalanine being oxidized to tyrosine, to dopa (dihydroxyphenylalanine), dopa is transformed first into dopamine and then into norepinephrine with the enzyme tyrosinase catalyzing the intermediate reaction. Other functions of organic copper include: vascular integrity, purine excretion, hemoglobin stability, gas exchange, digestive enzymes, myelin maintenance, visual accommodation, fertility, adrenal hormone synthesis, and connective tissue integrity.

Excess quantities of so-called vitamin C can reduce the pH of the urine to as low as 4.0, and cause the precipitation of large quantities of urates, increasing incidence of kidney stones. It appears that excessive doses of ascorbic acid interfere with purine metabolism, increasing the risk of gout. There is some evidence that high doses of ascorbic acid reduce fertility in some women. Large doses of ascorbic acid and vitamin E are undesirable as they may reduce availability of vitamin A. Ascorbic acid increases intestinal peristalsis and may produce diarrhea. This laxative action, along with its acidifying effect, may be the only benefit ascorbic acid has against a cold. High doses of so-called vitamin C, as popularly used in home remedies for the common cold, destroy substantial amounts of vitamin B12. Ascorbic acid in quantities in excess of a half a gram will destroy 50% to 95% of B12 content in food. Megadoses of phony vitamin C may produce B12 deficiency by destroying the cobalamins during transport through the digestive tract, and possibly also in the tissues. No further discussion should be needed to establish the lack of wisdom leading to taking or prescribing ascorbic acid.

11.12 VITAMIN E

The chemical name for vitamin E is tocopherol, which is derived from the Greek tokos (childbirth) and pherin (to bear). The ending ol is the chemical suffix to denote an alcohol. The name tocopherol was bestowed on this vitamin in 1938. Vitamin E is a fat-soluble vitamin, as are vitamins A, D, and K. Fat-soluble vitamins can be stored in the body whereas water-solubles cannot. Fat-soluble vitamins are excreted chiefly by the fecal pathway versus the urinary pathway for the water-soluble vitamins. They are absorbed along with dietary fats, and conditions of extremely low fat intake or impaired uptake of fats will also interfere with their absorption. Antibiotics and certain other drugs, as well as certain disease states such as malabsorption syndromes, decrease the absorption of fat-soluble vitamins from the intestinal tract.

One of the primary biological functions of vitamin E relates to its role as an antioxidant. Vitamin E's antioxidant ability is enhanced by selenium. This synergistic effect preserves membranes from destruction by oxidation products and especially retards breakdown (hemolysis) of red blood cells. Vitamin E can reduce the toxic effects of mercury. Vitamin E is able to reduce the chromosomal breakage caused by mercury. Vitamin E also has sulfhydryl-protective activity. When vitamin E is involved in reducing the toxic effects of mercury, there is less available for normal metabolic functions.

Vitamin E protects the enzymes G6P, G6PD, GSH-Px and glutathione reductase from auto-oxidative damage and at the same time protects the biological supply of glutathione. The presence of mercury in combination with a vitamin E deficiency has the potential of producing toxic reactions in the body. The primary lymphoid organs are major targets of selenium and vitamin E dietary deficiencies, contributing to impaired immune function.

Vitamin E is not a cofactor, such as the trace elements manganese, copper, zinc, and selenium, it is an independent antioxidant that has its main effect in the lipid phase of the cell, protecting the membrane from auto-oxidative damage. A distinguishing feature of cystic fibrosis is fat malabsorption. Cystic fibrosis patients usually have a deficiency of vitamin E. Patients with G6PD deficiency have less capacity to generate reduced glutathione, and their red blood cells have a greatly reduced half-life. Vitamin E supplementation can decrease the level of neonatal hyperbilirubinemia due to G6PD deficiency. Mercury inhibits G6PD activity and also reduces available vitamin E.

Alpha and delta tocopherols or mixed tocopherols are not vitamin E. They are merely antioxidant or protective agents (banana peel) for the more complex parts of the vitamin E assemblage (xanthine, phospholipids, lipositols, sex hormone precursors)--see functional architecture of vitamin E. There are seven tocopherols in all, alpha being one of them. The government has decided to rate any vitamin E product according to the amount of alpha tocopherol it contains. The tocopherols are antioxidants, just like ascorbic acid. They are a part of the E complex, but they are not the active ingredient. Their function is to preserve the active factor. Tocopherols are measured on the basis of cell growth.

The entire E complex contains the polyunsaturated fatty acids. (Vitamin F) Vitamin F is part of the E complex. Also, vitamins A and K. There are some forms of vitamin D and manganese in the E complex, as well. The traditional source for vitamin E is vegetable oils. But the best quality vitamin E comes from vegetables. Lettuce is one of the best sources and the pea plant is equally as good. If you take any laboratory-chelated products, your body has to do some combining to make them work. But if you take your minerals the way nature provides them, combined with protein and vitamins through the growing process of plants, they are bound to work. For example, we could take vitamin E and separate the tocopherols from the manganese. Putting them back together in the laboratory--chelating them artificially--we would find they wouldn't work like they did before you took them apart. If you take a watch apart piece by piece and analyze it and then throw all the parts back together, it isn't going to work! Although it is the same watch, it has now lost its function. A nutritional concentrate is an organized mechanism that has a function. If you break the function by separating the parts, the body can no longer use the mechanism. High potency vitamin E is made this way. Tocopherols are broken down and separated from oils--the natural tocopherols are separated from vegetable oil.

Then they are packed into a gelatin perle, and you have a high potency vitamin E product--a 220 or 500 International Unit (IU) Vitamin E product. That means 200 or 500 International Units) of alpha tocohpherol--the potency of the product is measured by the amount of alpha tocopherol it contains. Although the label of such a product will say "all natural" or "organic""--it was taken from oil, a natural source--it's still not food. It's an incomplete product.

The Vitamin E2fraction (prostaglandin) is found in beef chromatin material, which is glandular substance. It's a food factor that will stop a heart cramp or a heart pain, usually within about ten minutes. Most cramps are due to calcium deficiency. So E2 gets calcium into the area to relax the muscle. It can be used to relieve any muscle cramp, but it's especially effective to relax cramps in the muscles of the heart. Natural forms of vitamin E complex lose up to 99% of their potency when separated from their natural synergists. These synergists include tannins, fatty acids (vitamin F complex), phospholipids, and other synergistic factors. Chemically purified vitamin E (tocopherols), in high unit doses, reverses its effect and produces the same symptoms as a deficiency (including bone decalcification).

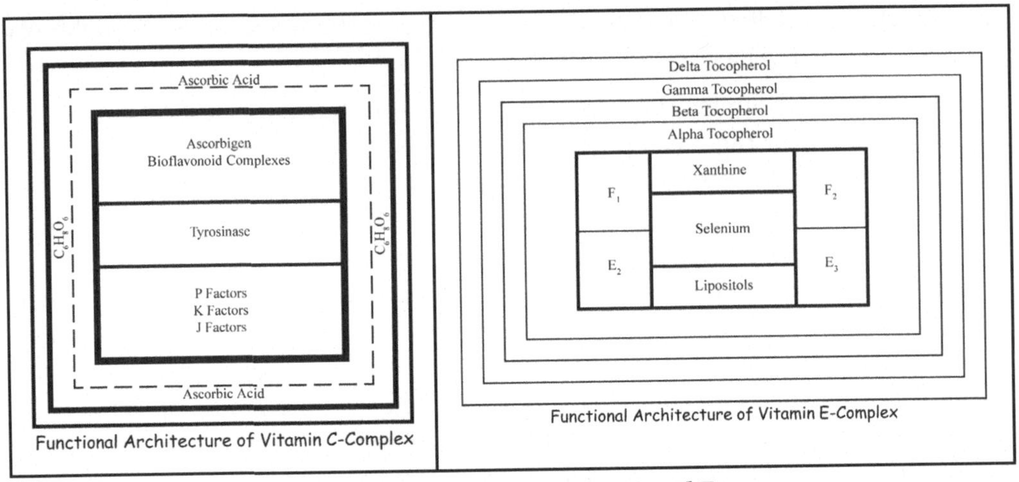

Fig Architecture of Vitamin C and E

Xanthine increases the antioxidative function of the tocopherols as much as 50%. Xanthine plays an important role in the metabolism of purines (nucleoproteins) within the cell. In one instance, hypoxanthine is transformed biochemically to xanthine with the metabolic fate proceeding to uric acid in two forms; enol and keto. Intracellular utilization of purines is dependent upon the said contribution. Lipositols are essential to cholesterol metabolism in order that cholesterol be altered to a biochemical acceptable to the synthesis of adrenal and sex hormones. F1 is the active form of unsaturated fatty acids. Vitamin F restores and maintains Calcium in tissue fluids and is anticarcinogenic. F^2 is part of the phospholipid complex in nerve structures; particularly the myelin sheath. It also promotes tissue repair and enhances the benefits of the antiarthritic Wulzen factor. F2 could very well be the missing link in pernicious anemia.

E2 is a part of E complex that has been compared to nitroglycerine in its function of relieving angina-type pains--the hypoxic spasm of the heart muscle. E2 reduces pain because it increases the oxygen conserving factor of the blood. The E3 component of Vitamin E complex is a precursor of sex hormones. Also, the E3 factor is similar to the cabbage juice Vitamin U which was developed and found useful in peptic ulcer syndrome. E3 is a powerful tissue rebuilder. In turn, E3 is considered as an important factor in the natural therapy of G.I. problems.

Drugs in five major groups have been shown either to function as vitamin B6 antagonists, or to increase the turnover of vitamin B6 in the body. Drugs in ten major groups have been shown to affect the absorption of folic acid; to act as folate antagonists; or to increase the turnover or loss of folate from the body. Drugs in four groups have been shown to affect the absorption of vitamin B12. Depletion of one vitamin can affect the requirement for another vitamin. Mercury can cause similar deficiencies or metabolic changes in the way the body handles these same three vitamins.

While man can isolate from foods known factors comprising about 99% plus of total weight, we cannot maintain normal nutrition or “life” by consuming isolated factors. Nature still withholds that “something” that sustains “life.” One of the deceptive practices of manufacturers of pseudo-nutrients is to label crystalline-pure and synthetic vitamin fractions as COMPLEXES. The meaning of complex is simply: the sum or combination of various things. Another, but more precise, definition is: Consisting of various connected or interwoven parts; composite. As a result, the consumer buys a “vitamin” product that says on the label: “Vitamins From Natural Sources (usually synthetic, from corn sugar and coal tar) with Bioflavonoids or Rutin.” Legally, this can be called a “Vitamin Complex” but it is really no more than a mixture of two or more dead, refined chemicals--not a required for life, “biochemical.” Remember, the human anatomy is composed of 100 trillion cells that need 75+ nutrients every day and, in turn, it can require six to eight months or more to overcome a nutritional deficiency. If you try and fool mother nature, she’ll make a fool out of you.

11.13 SYNTHETIC VITAMIN

A synthetic vitamin is created by a scientist in a laboratory when he/she has reconstructed the exact molecular structure of a vitamin by putting together or chemically combining molecules from other know sources. The synthetic vitamin has no synergists, no enzymes or co-enzymes, minerals, mineral activators, or co-vitamin helpers. It is a chemically pure vitamin but is, conformationally, a mirror image of the real thing. Legally it is not necessary to give the source from which the synthetic chemical is derived (coal-tar)

Item	If Source is	It is
Vitamin A	Fish oils	Natural
	Acetate	Synthetic
	Palmitate	Synthetic
	Lemongrass	Co-Natural
Vitamin B Complex		
Vitamin B^1 (Thiamine)	Yeast (unenriched)	Natural
	Thiamine mononitrate	Synthetic
	Thiamine hydrochloride	Synthetic
Vitamin B^2 (riboflavin)	Yeast (unenriched)	Natural
Pantothenic Acid	Yeast (unenriched) rice bran, liver	Natural
	Cal panto	Synthetic

Vitamin B3 (Niacin)	Yeast (unenriched), or bran	Natural
	Niacinamide	Co-Natural
Vitamin B6 (Pyridoxine)	Yeast (unenriched)	Natural
	Pyridoxine hydrochloride	Synthetic
Vitamin B^{12}	Liver or Yeast (unenriched)	Natural
	Cobalamin from fermentation	Natural
	Cyano cobalamin	Co-Natural/Crystalline
	Cyano cobalamin	Co-Natural/Crystalline
PABA	Yeast (unenriched)	Natural
	Para-aminobenzoic acid	Synthetic
Folic Acid	Yeast (unenriched) or Liver	Natural
	Pteroylglutamic acid	Synthetic
Inositol	Soy beans	Natural
	Reduced from corn	Co-Natural
Choline	Soy beans	Natural
	Choline chloride	Synthetic
	Choline bitartrate	Synthetic
Biotin	Liver	Natural
	d-Biotin	Synthetic
Vitamin B15 (Pangamic acid)	Bran-apricot kernels	Natural
	Pangamic acid	Synthetic
Vitamin C-Complex		
Vitamin C	Buckwheat leaf, peppers, tomatoes, citrus, rose hips	Natural
	Acerola berries	Co-Natural
	Ascorbic acid	Synthetic
Vitamin D-Complex		
Vitamin D	Fish oil	Natural
	Irradiated ergosterol	Co-Natural
	Calciferol	Synthetic
Vitamin E-Complex		
Vitamin E	Wheat germ oil	Natural
	dl-alpha tocopherol	Synthetic
	d-alpha tocopherol acetate	Synthetic
	d-alpha tocopherol succinate	Synthetic
Vitamin F (Unsaturated fatty acids)	Flax oil, pumpkin seeds, walnuts, soybeans	Natural
Vitamin K	Alfalfa, Cabbage	Natural
	Menadione	Synthetic

REFERENCES

Fekete, S.; Guillarme, D.; Sandra, P.; Sandra, K. Anal. Chem. 2016, 88, 480-507 and references therein.

Yu, L.; Remmele, R. L.; He, N. Rapid Commun. Mass Spectrom. 2006, 20, 3674-3680.

Zhang, Z.; Pan, H.; Chen, X. Mass Spectrom. Rev. 2009, 28, 147-176.

Fodor, S.; Zhang, Z, Anal. Biochem. 2006, 2, 282-290.

Srzentic, K.; Fornelli, L.; Laskay, U.; Monod, M.; Beck, A.; Ayoub, D.; Tsybin, Y. O. Anal. Chem. 2014, 86, 9945-9953.

Siemiatkoski, J.; Lyubarskaya, Y.; Houde, D.; Tep, S.; Mhatre, R. Carbohydr. Res. 2006, 341, 410-419.

Reiding, K. R.; Blank, D.; Kuijper, D. M.; Deelder, A. M.; Wuhrer, M. Anal. Chem. 2014, 86, 5784-5793.

INDEX

9 789390 013715

Printed at Repro India Ltd.